6,000

OF

E, COLO.

27. Gas Works.
28. Denver and Rio Grande R. R. Depot.
29. Denver and Rio Grande R. R. Round House.
30. Door and Sash Warehouse and Lumber Yard, J. D. Griffiths, Prop.
31. Planing Mill and Shingle Manuf'y, W. D. Scott, Prop.
32. Tabor Milling Company's Works.
33. The Leadville Daily and Weekly Democrat, Democrat Printing Co., Publishers
34. The Leadville Daily and Weekly Herald, Herald Printing Co., Publishers
35. Chronicle.
36. Leadville Gold and Silver Mill Co.

A—Baptist Church.
B—Congregational Church.
C—Episcopal
D—Methodist
E—Presbyterian
F—Roman Catholic Church of Enunciation
G— " " " of Sacred Heart
H—A. Methodist Episcopal Church.
K—St. Vincent Hospital

Ann & Dulane –

To my very special

Cousins – Love

Betty

December 1979

THE MINERAL BELT

Volume I — Old South Park — Denver to Leadville

DAVID S. DIGERNESS COLLECTION

COLORADO & SOUTHERN locomotive Number 37 simmered to herself in Denver in 1915. She was a veteran of the "Old South Park."

THE MINERAL BELT

Volume I — Old South Park — Denver to Leadville

An Illustrated History

Featuring the Denver, South Park & Pacific Railroad
And the Gold-and-Silver Mining Industry of the
Fabulous Mineral Belt of Colorado

BY DAVID S. DIGERNESS

And the Editors of Sundance Books
With Special Photographic Contributions
From the Collections of
John W. Maxwell
and
Richard A. Ronzio

SUNDANCE
Books

THE MINERAL BELT

An Illustrated History

By David S. Digerness
And the Editors of Sundance Books

Published by Sundance Publications, Limited,
Silverton, Colorado

Graphical Presentation and Printing by
Sundance Publications, Ltd., Silverton, Colorado

Binding by Mountain States Bindery,
Salt Lake City, Utah

Editor-in-Chief — Russ Collman
Editor — Kay Dolliver
Production Manager — Dell A. McCoy
Color Separations — Steven J. Meyers
Printing — Dave Curnutt

ISBN — 0-913582-20-4

First Printing — November, 1977

Acknowledgements

The author would like to give special recognition to the following people who gave their personal encouragement for completion of this book.

The two people who gave me the greatest drive to push *The Mineral Belt* to completion were the late Mellie Bowman White of Georgetown and the late M. C. (Mac) Poor of Denver. They gave me the incentive I needed to complete the compilation. They also helped me realize that my desire for the perfection of the work was causing unnecessary delay, so, with this in mind, the manuscript was brought to the final stage of preparation.

A further incentive was my intriguing collection of original photographs of Colorado, which would enable me to illustrate the books with authentic early-day views. This then was enhanced by adding still other rare color photographs from John Maxwell, and photographs from the collections of the State Historical Society of Colorado and the Denver Public Library, Western History Department. In addition, pictures from the files of my free-lance color photography hobby over the years add a present-day look.

The name of Richard A. Ronzio stands out as the foremost outside contributor to *The Mineral Belt* — providing an assortment of historic sepiatone views second only to my own personal selection of illustrations for the book. Mr. Ronzio's contribution consists of some of the rarest early-day Mineral Belt railroad and mining views yet discovered. Without his gracious assistance in this regard, *The Mineral Belt* would be a much poorer book.

The assistance of many other people, too numerous to list here, contributed greatly. I wish to thank all of them, particularly my parents, who gave their full support through the years of compilation. I especially want to recognize my mother, who gave unsparingly of her time in typing and retyping the manuscript and photo captions, as well as for her authoritative contributions.

I would welcome hearing from persons who may detect imperfections or errors, and from anyone who has some of Jesse S. Randall's missing personalized material known to exist. Hopefully, these items will turn up in someone's private collection. Also, any old photographs that might be used in a possible later revised edition would be welcomed.

David S. Digerness
Denver, 1977

THE MINERAL BELT AREA COVERED IN THIS VOLUME.

Introduction

AS THE NAME IMPLIES, "The Mineral Belt" refers to a large geographical area of mineral wealth — centered at Georgetown, directly west of Denver in the great Front Range of the Colorado Rockies. Technically referred to by geologists as the Main Belt or Leadville-Boulder Belt, and more popularly known as the Front Range Mineral Belt, this region is a continuous mineral belt which runs from the proximity of the town of Boulder to Leadville, in a direction nearly due southwest-northeast, about 85 miles in length and 5 to 12 miles in width — passing directly through the Georgetown mining area.

The Mineral Belt covers areas in Boulder, Gilpin, Clear Creek, Summit, Park and Lake counties, as well as some areas outside of this, such as Denver, Arapahoe, Jefferson, Eagle, Chaffee, Gunnison and Pitkin counties. The counties of Chaffee, Gunnison, Pitkin and Eagle are included in the total extent of what is known as the General Belt, as measured from Boulder to the area north of the town of Gunnison — in proximity to the ghost towns of Ruby and Irwin — a distance of 130 miles. This region also is known as the great Rocky Mountain Gold-and-Silver Belt, which includes the Aspen mining area. To the northeast of Aspen, a part of the belt passes through Lenado and into the Frying Pan district. Some distance northeast of the Frying Pan region is the Redcliff district, which is only slightly northwest of the main Leadville-Boulder Belt at Leadville. To the southwest of Aspen, the belt is traceable through Ashcroft and other mining localities across the Elk Mountain Range into Gunnison County, where it includes the Ruby, Irwin, Crystal and other neighboring mining districts. Also included is the area of a continuous north-south belt to the southeast, which embraces the Tincup and Monarch districts. The belt which passes through Aspen — if considered to extend from the Frying Pan country to the Gunnison region — is about 45 miles long and 5 to 6 miles wide, and is a northeastward-trending mineral belt, slightly offset from the Main Belt. This area outside the Main Belt is included in the first two volumes along with the southern area of the Main Belt.

The first two volumes in *The Mineral Belt* series are subtitled *Old South Park Railroads* and present the most comprehensive word-and-picture story of the Denver, South Park and Pacific narrow-gauge system ever attempted. Also included is extensive coverage of the mining industry served by the railway, as well as other inter-related rail lines of the region described above — and the lay of this land is one of the most beautiful in North America.

Many of the facts for this exciting and informative saga have been taken from the travel journals and news material gathered by Jesse Summers Randall, the founder and publisher of *The Georgetown Courier.* All of this material — at times provided by Randall's associates and members of his family — relates to the amazing history of The Mineral Belt. *The Mineral Belt* series presents the priceless records and writings of one of Colorado's extraordinary early-day newspapermen as a comprehensive history of the state's Front Range Mineral Belt. These wondrously revealing and colorful volumes cover the entire dramatic story of railroading and mining in this region — spanning over 100 years — to make this a truly outstanding pictorial documentary. Throughout *The Mineral Belt* series you will gain a deeper understanding of one of North America's most splendiferous regions of natural wonder. A detailed exposition of the geological forces which created and shaped the mountains of The Mineral Belt and the full history of man's involvement will provide you with a new appreciation of the men and women who made the struggle that provides herein. . .

Drama. . .Beauty. . .and History

Years after Jesse Randall passed away, personal records came to light, including files of *The Georgetown Courier*, letters and journals. Also included were scrapbooks containing clippings, photographs and related materials, and other of his writings, as well as the choicest part of his personal library of rare books and early maps. His daughter, Pearl Randall Sidney, kept much of this material stored away along with other priceless items saved by my mother. Preserved as it has been for many years, this material is an important treasure of Colorado history. It covers the exciting pioneer years of The Mineral Belt — as well as subsequent years, when the impact of the wealth flowing from its mines was being felt in Colorado and in the United States as a whole — the financial disaster of 1873, the demonetizing of silver in 1893, the closing of the mines, and the long wait for the revival of silver and mining as a whole. The story is presented from the viewpoint of Jesse Randall, as he actually lived it — from the time he arrived in The Mineral Belt and continuing throughout his career with *The Georgetown Courier.* It includes his many travels by railroad and other modes of transportation, the region's mining industry, the building of the railroads, the towns and mining camps, and the annals of the '59ers (narratives of their deeds and adventures, and observations thereof). Jesse's travels radiated from Georgetown, Colorado — then known as the *Silver Queen of the Rockies* — in the heart of The Mineral Belt. These journeys provide the main focal point for relating the stories about the many fascinating places in this region — both past and present. The news coverage of the day in part consisted of what was happening in the mining camps and told about the people, the mines and mining activities, the latest in new mill construction, mining methods and smelters, and personalities connected with the above. It also reported the building progress in the new towns and camps throughout the area, the transportation news, (such as the coming of the railroads), news of proposed new enterprises and progress reports on existing ones. Some detailed statements of production from the many mines and mining districts are included, along with new discoveries, and of course, items from Jesse's famous social column. This column kept everyone informed about everybody else's business, and along with obituaries, proved to be valuable historical information.

Jesse Randall's eyewitness reports presented in the *Courier's* columns and in other records are impressive — written in his precise, dignified, carefully planned and executed style. Descriptions of the hectic behind-the-scenes activity in gathering the news before press day — Thursday of each week — are set forth, telling of the work required to gather stories from all over The Mineral Belt. It should be pointed out that the newspaper was always delivered by mail, even in Georgetown, so it was necessary to get each issue to the post office by no later than Friday night of each week. As one reads Jesse's stories, his character takes on new dimensions, and you begin to understand what an extraordinary pioneer newspaper editor he really was, even under the great stress of having prominent people against him and his ideals.

In connection with minerals, the name of Jesse S. Randall was widely known. After he and his father arrived in Georgetown, they both familiarized themselves with geology, mineralogy and mining, and established a reputation for their knowledge in these sciences. They gathered together a very fine collection of minerals and Jesse printed one of the first books ever published on the minerals of Colorado. A special selection of Colorado minerals from his famous collection — obtained from all over the world — was exhibited at the World's Columbian Exposition held in Chicago during 1893.

With Georgetown as the homebase for our story, considerable attention is given to this settlement and the immediate surrounding area, and the saga of The Mineral Belt always radiates from this mining town on Clear Creek.

The Mineral Belt's railroads played a very important role in Colorado's transportation, and during their time they were used extensively by Jesse in his travels while gathering the news in this vast area. There were many railroads which served the region, but only one main one which served the Clear Creek drainage. This will be covered in the third volume of *The Mineral Belt*, subtitled *Canyons and Valleys of Clear Creek*. During the heyday of this line — called the Colorado Central when first built — there were four trains in each direction every Sunday during the summer season, filled with tourists and picnickers out for the thrill of riding over the world-famous Georgetown Loop, a marvel of engineering skill in those days.

My grandmother, Pearl Randall Sidney, who was born in Georgetown in 1875, remarked to me that she saw the high iron bridge over Clear Creek while it was being constructed, and regretted having witnessed its dismantling as well. This bridge was called the Devil's Gate Trestle and made up the most spectacular part of the loop. It stood during some of the most exciting years The Mineral Belt ever experienced — the great narrow-gauge railroad era.

The automobile and motor truck have taken the place of the train, and there now is a broad highway leading to the new Eisenhower Tunnel — opened in March of 1973 — piercing the Continental Divide west of Georgetown. The original highway route went over Loveland Pass, a pass named for W. A. H. Loveland (named somewhat ironically it seems, in the light of what has happened). He headed the company which built the first railroad into The Mineral Belt, connecting the region to the outside world. This narrow-gauge line faithfully served these mountains for many years — as Volume III of *The Mineral Belt* will bring out in detail.

As a native Coloradan and member of the fifth generation from Georgetown, my interest in the heritage of this region built gradually as I got older. Many of my family's treasures were disposed of before I realized what was happening. Then — all at once — I seemed to gain a great appreciation for what was being lost, and this book was started shortly thereafter, in 1954. My interest came mainly from being a native of Georgetown, and the fact that my great grandfather on my mother's side and grandfather on my father's side were newspaper editors who owned their own plants. The former, of course, was Jesse Randall, and his newspaper was located in the mining country of Clear Creek County, while the latter's plant was located in the farming country of Adams County, in the town of Hudson. In addition, uncles on both sides had newspapers at other towns, both in the Rockies and out on the High Plains.

Photography caught my attention at an early age and led to my overall interest and appreciation of this art-science. The Mineral Belt had more railroads serving it than a railfan photographer could ever dream of, and part of my interest came from watching the fascinating narrow-gauge trains running up and down Clear Creek Canyon as a youngster.

What was it like to ride the highest standard-adhesion railroad in the world, climb over the great Continental Divide on a narrow-gauge train — or, if you like, by standard-gauge, to travel through the Divide instead of over it — or to travel into the many mining towns the early-day railroad served?

Much of *The Mineral Belt* is devoted to the coming of the iron trail and its fabulous iron horses, and its subsequent history. Considerable insight into this will be acquired by reading all the volumes of this series.

Some of the oldest ghost towns of Colorado are found in The Mineral Belt. I have always been intrigued by ghost towns and have traveled many trails to photograph them, and many times I have wondered what it was like living in these settlements during their heyday, and why they were located where they were — and I have often looked for traces of other towns, which, often as not, have vanished from the face of the Earth.

Many of the ghost towns of the area will be covered in this publication — with each volume providing information about various localities. Visitors to Colorado, as well as residents, no doubt have seen many of the historic landmarks of the area: the numerous abandoned mine buildings, mine dumps, tailing piles, foundations of large mills, abandoned railroad grades, old cemeteries, etc., and would like to know more about the history of these ghosts of a bygone era. You will find it in *The Mineral Belt*, along with interesting narratives.

I have been interested in rocks ever since my grandfather showed my brother Ted and me samples from his mine when the two of us were children. Rockhounds, mine and Western history buffs will find a bonanza of information about Colorado's mineral wealth in these volumes.

The mineral resources of Colorado are known to be enormous in quantity, and yet further discoveries are constantly being made. At first, the gold seekers passed over preciously rich metallic-silver deposits — with scarcely a sideward glance — and as prospectors beat paths to all the different early-day mining locations, they passed by many other valuable minerals, which as time passed, came into their own. Molybdenum is one — found in abundance in The Mineral Belt — an alloy whose value was not known until World War I, when the United States was looking for fine steels. Ever since then it has been used to produce high-quality toughened steel.

Atomic-age metals, or radioactive minerals — such as uranium carnotite, pitchblend and vanadium — came into their own during World War II, which resulted in the atomic bomb and brought an end to the war. In the 1950's another mining boom was on, and prospectors with Geiger counters or scintillators could be seen combing the areas of western Colorado, Utah and Wyoming. This time, instead of burros, they used Jeeps and even aircraft armed with the sensitive devices required to detect radioactivity. Some new discoveries did occur in areas of The Mineral Belt, but not in abundance and always it seemed to be in hard granite. The Colorado Plateau, in far-western Colorado, outside The Mineral Belt, proved to be one of the richest areas of discovery. This was the area where uranium was first mined in Colorado. Part of the first shipment of uranium minerals going to Madame Curie in 1898 came from Uravan, Colorado, and also included minerals from Central City. Madame Curie's experiments with this material led to the first extraction of radium. For a 10-year period — between 1910 and 1920 — this area was the world's chief supplier of uranium. Thorium, one of the heavy elements, is another atomic-age metal, and according to its proponents, it has proved to be the best fuel for the thermal reactor of the atomic furnace. Uses have been developed for heat-resisting metals in missile nose-cone heat shields and other products. The high-temperature metal, columbium, has come into its own as one of the metals in this field.

According to figures compiled by the U. S. Bureau of Mines, the year 1972 saw the value of Colorado's mineral production rise to its highest mark ever, amounting to $420,000,000 — and every year since then has seen increases in production.

Using 1972 figures, I want to point out that mineral fuels, primarily petroleum and natural gas, accounted for 7 of the 34 mineral commodities produced in the state. Twelve metals and 15 non-metals also were mined. Petroleum and natural gas — found in large quantities in the Plateau Province oil fields of Colorado — represented 75 percent of this mineral fuel production. Most of the 12 metals produced in Colorado were mined in The Mineral Belt districts, which contain the largest mineralized section of the state — with only five other areas accounting for major mineral deposits: Creede in Mineral County; Cripple Creek in Teller County; Silver Cliff in Custer County; the Telluride, Ouray and Silverton area in San Miguel, Ouray and San Juan Counties; and the Western Slope vanadium-uranium districts. With these other districts, the collective production was valued at $165,700,000.00 in 1972, or 39 percent of the total mineral value of the state.

Based on dollar value, the three leading minerals in the metallic group were molybdenum, zinc and uranium. Colorado maintained its position as the leading molybdenum (and its by-product, tin) producer in the nation, which it has held for several years. It moved up to second place in zinc output and ranked third in uranium production. Zinc and lead are scattered throughout many of the mining districts of The Mineral Belt as primary ore, and as a minor constituent or co-product of base- and precious-metal (gold-and-silver) mineralization. Zinc production for the year 1972 was 68,800 tons, with a value estimated at $24-million. A jump in lead production was recorded for the year. The value of this lead was set at $10-million. Colorado's mines dug slightly more than 34-million tons of lead ore in 1972.

Although gold and silver have not been mined extensively in Colorado since Federal government mine closures during World War II, both precious metals showed high increases in value. The increases were a direct result of the sharp price advance of free-market gold in 1972, boosting the average price for the year more than 40 percent. This put the price at about $58.25 per ounce, compared with $41.25 just the year before. The spiraling price of free-market gold, caused by monetary fears, sent gold up to $114.15 per ounce on August 8, 1973, and it was going steadily higher. As a result, hardrock miners renewed their interest in exploration for precious metals, and there was a good possibility that many of the state's old mines would be reopened. However, most prospective small-mine operators of today do not realize the cost and time involved in reopening old mines or starting new ones. Many problems face mine owners, such as higher costs of equipment, labor, transportation, milling and smelting, as well as new mine-safety rules, pollution regulations and a lack of experienced miners. There are shortages of many items essential to mining. Therefore, it seems more than likely that it will be the large-mine operators who will benefit from the rising prices; the small operator may benefit in time, if the price of gold gets high enough and inflation is kept under control.

Motivated by the exemption of commercial-grade silver from the price ceiling of just under $1.62 per ounce, silver production and value increased to $7-million, derived from six-million troy ounces of production in 1972.

The majority of the state's molybdenum ore is mined in a portion of the region covered in our story — in the central part of the northeast-southwest-trending Front Range Mineral Belt. In 1972, developed primary deposits at the Climax and Urad mines, of American Metals Climax, Incorporated, provided almost all of Colorado's production of this metal. The operating capacity of the Climax mine and milling circuit atop Fremont Pass — partly in Summit County and partly in Lake County — is about 43,000 tons of molybdenum ore per day. The Urad mine was a short-term venture of the company, with the ore deposit exhausted as of November 24, 1974, at which time production ceased. The Urad mine produced approximately 5,000 tons of ore per day. It was the largest mine in Clear Creek County and, since 1967, has yielded 14-million tons of ore, which produced approximately 49-million pounds of molybdenum concentrate. The company's Henderson mine, just west of the Urad mine, is now in full operation — mining a tremendous molybdenum-ore deposit, which will take many years of work to exhaust. It was in development for ten years and was engineered to the most up-to-date techniques. This mine is covered in Volume IV *The Mineral Belt.*

Uranium mining is conducted today in Uravan's mineralized area in southwestern Colorado and from pitchblend ores in The Mineral Belt.

In 1972, mine production of non-metals — including sand, gravel, feldspar, clays and cement — was the highest in history. Cement production during that year, for the first time in the history of mineral production, was the leading non-metallic commodity in Colorado. The substantial increase was attributed to higher cement prices and a very active construction industry, and has continued to break records each year since then.

Non-metallic minerals have been important to the Colorado scene ever since their first use by early settlers. One of the most important uses was clay for brick production utilized in the construction of fireproof buildings, as well as for firebrick for coke ovens, fireboxes, smelters, etc.

Today, industrial minerals from Colorado — such as limestone for cement, steel production and sugar-making (one of the biggest uses); gypsum for sheetrock or wallboard; sand, gravel and dimension rock for construction; bentonite for sealing purposes, etc. — all have made a great mark on the area's economy. Coal — another non-metallic mineral — was one of the most important commodities of all when the state was new, and now is coming back into its own after slacking off in the 1950's. Many miles of railroad were built just to tap the reserves of coal, as was the case when trackage was laid to reach gold-and-silver mines during the early mining booms.

Present use of coal has caused railroads to run the largest number of unit trains ever in the history of the state. Also, today's use of natural gas and petroleum has made these the biggest fuel resources for energy — even though coal has far more potential for the future. Natural gas has been piped into The Mineral Belt from the southwest to points such as Leadville and the Climax mining-and-milling circuit atop Fremont Pass. It also has come into Clear Creek County to Idaho Springs, Georgetown, Empire and its great neighbor to the west, the Henderson mine.

At the time when this book was undergoing final preparation — during December of 1974 — after extreme spiraling of the price of free-market gold for the previous two years, gold reached a record world price of $201.40 per ounce in Paris and edged close to the $200.00 mark in other European markets. U. S. legalization of the private ownership of gold by individual citizens, which allowed them to buy and sell at will, took effect the same day. There is no guarantee as to the value of gold bullion in a fluctuating market, and an expected rush to buy did not occur, causing prices to drop.

On January 6, 1975, the U. S. Treasury tried auctioning off two-million ounces of gold from its 276-million-ounce stockpile. The sale involved 400-ounce bars and was a failure, for Americans did not respond to the high market prices.

Some of the mines in The Mineral Belt have once again been opened for production. In Gilpin County, mines such as the Cyclops and the Bonanza have been reopened; in Lake County, the Black Cloud at Leadville, with a production of around 1,000 tons of rich ore per day, has reopened; and in Clear Creek County, the John Paul Jones has resumed operations — just to mention a few. The small miners have many problems — as already mentioned — but, with the right to own gold, the new high gold prices, and the rising price of silver and other minerals, it appears that the unmined ore deposits can again become profitable to mine. Mills and smelters will have to make some kind of a comeback if the small miner is ever going to come to the front, however.

In the meantime, the so-called "energy shortage" has been foisted upon us, affecting everyone and our national economy as a whole. It should be pointed out that more self-sustaining facilities, such as the Public Service Company's Cabin Creek hydroelectric plant — located near Georgetown — could be built with a minimum amount of harm to the environment. (This plant will be expounded upon in Volume IV of *The Mineral Belt*.) Another possible means of producing energy, not to be forgotten, is the harnessing of wind in perpetual high-wind centers. The high plains, plateaus and mountain ridges of the West could easily provide a ready supply of energy to help produce electric power for the region.

The re-establishment of electric-powered mass transit in Denver, as well as in other towns, could partially eliminate the automobile's harmful air pollution from the areas of the greatest population density, and in turn, help eliminate the energy shortage. Solar energy should be harnessed to its fullest capacity as rapidly as possible. This is one of our greatest sources of energy, but it has not been utilized except in space technology.

Air, water, trash and noise pollution are all problems we must overcome after 100 years of settlement in Colorado, but with population mounting at an alarming rate, and over-population of the state (at least in certain areas), the whole environment is in danger. Damage to our wildlife already has reached shocking proportions.

The section of the book dealing with The Mineral Belt of today includes tourism and year-'round recreation, the rush for mountain property, the excellent historical preservation work being done by various groups, the great tunnels for transportation and water diversion under the Continental Divide, and the large mines presently being worked. This has been an era of accomplishment and exploitation; a wilderness has been tamed and its resources exploited by active, pragmatic — and sometimes greedy — men who followed visions of personal and national greatness. Unfortunately, many of Colorado's well-known historic monuments — masterpieces created by these men in the form of great architecturally designed buildings in Denver — have fallen under what is called today's "progress". Modern highways in The Mineral Belt often have destroyed landmarks along their rights-of-way. The elements and vandals — especially vandals — have taken other landmarks which had survived up until a few years ago. More respect and preservation interest by all — and especially by our lawmakers and big-business interests — are needed to help save some of our historic landmarks, ghost towns and endangered areas of living history. What a terrible price we pay for "progress" if we destroy our heritage and despoil our environment in the process.

My Great Uncle Neal — Jesse's son, "Neal" Cornelius Randall — pursued a great but unsuccessful campaign to save the Georgetown Loop railroad from abandonment. Today the "Georgetown Loop" itself is being reconstructed under the auspices of the State Historical Society of Colorado. In Gilpin County, the rail line which ran from Black Hawk to Central City also is being reconstructed.

It is my firm belief that we need not destroy our historic past to make way for the new. I realize that the present situation is just another transition in the process by which man constantly moves toward an ever-more-modern world, and not everything that is old can, or should, be saved. However, we should always ask if "more modern" is necessarily *better*. Beyond that, everyone should remember that as a tourist, one likes to see the historic landmarks of history — if they are to be found.

It has been gratifying to see the formation by our Federal government of the National Historic Sites Program, administered by the U. S. Department of Interior's National Park Service. This body has the power to declare areas to be set aside for this purpose, and one of the areas chosen as a National Historic Area is the Georgetown Loop Historic Mining Area. In 1974, the towns of Georgetown and Silver Plume were included in this area, and the whole district was designated the "Georgetown-Silver Plume National Historic District." (It is interesting to note that Georgetown was the first settlement in Colorado to establish a local historic preservation ordinance.)

In all, *The Mineral Belt* is dedicated to the memory of Jesse Summers Randall and his daughter, Pearl Randall Sidney, and commemorates the 100th anniversary of the founding of the *Georgetown Courier*.

David S. Digerness
Denver, 1977

A. E. MATHEWS SKETCH
DAVID S. DIGERNESS COLLECTION

GULCH MINING, a form of placer mining, was done in placers and gulches containing alluvial or glacial deposits. Here, the work was more simple than in lode mining, and the gold was more readily obtained.

CONTENTS

W. H. JACKSON PHOTO
DAVID S. DIGERNESS COLLECTION

THESE UNUSUAL uptilted sandstone rock formations are located near the mouth of the South Platte River, where it emerges from the Foothills. These formations jut out from the Earth's surface at different locations along the foothills of the Front Range, which run north and south, forming escarpments just west of the High Plains.

Chapter I

Lay of the Land and Its Geology

THE MINERAL BELT is an area of majestically beautiful mountain topography, flanked by the High Plains to the east and additional mountain ranges to the west. Visualize, if you will, the vast extent of this area — bounded by a line east of the mountains, running north and south through Denver, Longmont and Castle Rock, a distance of approximately 65 miles. Also visualize the area west of this line into the mountains, all the way to Leadville, and a line running just west of and along the Continental Divide from Leadville to just northwest of Boulder. Then include the area of the extension of the Main Mineral Belt and vicinity to the west and southwest of Leadville as far as a north-south line between U. S. Highway 50, a short distance west of Gunnison; and the new Interstate Highway 70, just west of the town of Glenwood Springs. The latter area is covered in this book because of its relation with the Main Belt.

The area of the greatest mineral discoveries has been traced from northern Boulder County south to Gilpin and Clear Creek Counties, and to the southwest through Summit, Park and Lake Counties. Rich mineral discoveries were made on both the western and eastern slopes of the Continental Divide, which runs along the great Sawatch Range and forms a dividing line between Gunnison, Lake and Chaffee counties. Silver was the predominate precious metal in this area. East of the Sawatch Range — in the southern portion of The Mineral Belt — is the Mosquito Range, which separates the mining camps of the Arkansas Valley from those of Alma and Fairplay in South Park. The Mosquito Range, joining the Sawatch Range in the vicinity of Mount Lincoln, was enormously productive. On its western slope were the world-renown carbonate deposits and veins of Leadville, immeasurably rich in silver and lead, and the gold veins and alluvial deposits of California Gulch. Gold, copper and lead were mined in the vicinity of the Mosquito range in addition to silver. To the south were the gold placer mines of South Park, and to the northwest, across the main range, were numerous argentiferous and galenous veins, as well as gold-producing gulches and placers which were worked for many years.

The Main Mineral Belt contains the headwaters of the mighty Arkansas, Platte and Colorado rivers and many of their tributaries, which drain to each side of the Continental Divide. The tributaries of both the South Platte and Arkansas drain to the Atlantic Ocean. These vast drainage systems have helped form the lay of the land. On the 14,000-foot peak of Mount Evans, southeast of Georgetown, the drainage of the northwest, north and east sides form part of the upper drainage of South Clear Creek and Bear Creek. North Clear Creek drains the Central City, Apex Mining Areas. Clear Creek and its affluents drain the middle portion of the Mineral Belt. The south drainage of Mount Evans flows into the tributaries of the North Fork of the South Platte River. Boulder Creek, farther north, along with other creeks such as the South Saint Vrain, drains the northern portion of The Mineral Belt with the help of their many drainage tributaries. Some of the principal streams of the eastern slope or Atlantic drainage — all tributaries of the South Platte River — include from north to south: the Saint Vrain, James, Left Hand, Boulder, Clear, Bear, Turkey, Elk, Deer and Geneva creeks and all their affluents. The last three are tributaries of the North Fork of the South Platte. The main drainage of the South Platte River has its headwaters in the mountains surrounding the South Park area in the southern portion of the Mineral Belt. All are part of the South Platte River drainage basin, which, in turn, is part of the vast Missouri River system. The streams are all mountain torrents of steep grade, and with the exceptions of Clear Creek and the South Platte River (the largest or main channels), the tributaries fall 300 or more feet per mile. The main channels of both Clear Creek and the South Platte River are rapid streams from 30 to 60 feet wide, having an average fall of about 85 feet per mile. Two distinct types of stream profile are well illustrated by the Soda Creek and Chicago Creek tributaries of Clear Creek.

Soda Creek heads at a comparatively low altitude and has a normal compounding curved profile. Chicago Creek heads in glacial cirques, and, in consequence, has a bench-and-scrap profile in its upper reaches. The Arkansas River heads at Fremont Pass above Leadville, and, with all its affluents, drains south out of the southern portion of the Mineral Belt to the Atlantic Ocean. All drainage west of the Continental Divide drains to the Pacific Ocean by affluents of the Colorado River.

When the area was first settled, much of it was heavily forested below timberline (which ranges from 11,250 to 12,000

DAVID S. DIGERNESS PHOTO

IN THIS VIEW, a mantle of newly fallen snow covers the contorted landscape of Red Rocks Park. These great slabs of red sandstone — along the Foothills of the Front Range — were tilted upward millions of years ago.

ON TOP of one of the great tilted slabs of red rock are two formations resembling lizard heads, or the heads of prehistoric monsters. Strangely enough, evidence of prehistoric life has been found in the area. This scene is in Red Rocks Park, one of the Denver Mountain Parks owned by the city and county of Denver.

DAVID S. DIGERNESS PHOTO

feet above sea level). There are areas, for example, southeast of Mount Evans, where trees are rather widely spaced and grassy grades are characteristic. Western yellow pine and Douglas fir grow in dry areas to altitudes somewhat over 8,500 feet. These trees are replaced on higher and moister ground by lodgepole pine and Engleman spruce, and near timberline, by mountain white pine. Larger areas have been burned over and are now covered with second-growth lodgepole and aspen. Many areas near the mining towns, and for miles around, were cut clear in the early days and also have been replaced in some of these areas by this same woodland cover, while some remain clear with only remnants of great forests of trees in evidence in the form of weathered stumps. In the summer, the aspen's light green, and, in autumn, its yellow and yellow-orange, contrast strongly with the dark green of the conifers and add much to the beauty of the mountain-sides. Above timberline there is an alpine flora comprising many species.

J. H. BROWN PHOTO
DAVID S. DIGERNESS COLLECTION

THE TWO FLAT-TOPPED MESAS in the distance are North Table Mountain (to the left) and South Table Mountain (to the right). Geologists believe that Clear Creek — flowing down through the mountains for millions of years — cut Table Mountain in half. In this view overlooking the High Plains (to the east), the town of Golden is shown nestled in the valley.

In consequence of the great range in elevation, the climate of the area is varied; Georgetown and Idaho Springs have a mean annual temperature of about 41°F. The temperature seldom rises above 80°F. in summer or falls lower than 5° below zero in winter. The climate in the higher country is more rigorous and the annual range in temperature greater. The valley towns have an annual recorded precipitation of 17.5 to 24.8 inches, 19.5 inches being the average. The precipitation on the higher peaks is much greater, and Mount Evans, as well as the high peaks along the Main Range, are the storm centers of the Front Range. April, May, June, July, and August are the months of greatest, and January that of least precipitation. During the summer of an average year there are showers almost every afternoon. The snowfall starts building up in the high country often about the middle of October, and a snowpack accumulates during the winter months until early spring when it starts melting, supplying water for streams year-around. The lower valleys and canyons are usually free from snow much of the winter, but snow lies on the highest peaks all but two to three months out of the year.

To get a better idea of the lay of the land of our story and the tremendous obstacles which faced the pioneers who traveled and settled this mountainous area, one should know something of its geology and thus know how it all came about.

Recorded History in the Rocks

The geological forces that created and shaped the mountains and surrounding vicinity of The Mineral Belt are a most interesting subject. Precious gold, silver and other minerals were created by the mountain-building forces and were the main attraction for penetration and settlement of the area; also, the magnificently beautiful topography is a result of all those forces. The rocks of the foothills and the plains are mostly of sedimentary origin. While the rocks of the mountains are mostly of igneous and metamorphic origin, formed more than a billion years ago in Precambrian time, some of them are more than a billion and a half years old. Their age has been determined by measuring the amount of decay of radioactive minerals, a process that takes place at a constant and determinable rate. Most of the Precambrian rocks have been so drastically changed by pressure and heat deep within the Earth that their original character now is obscured. Banded crystalline rocks called *gneiss* and micaceous rocks called *schist* originally were mostly sedimentary rocks, much like those along the foothills, but vastly older. Conspicuous light-colored ribbons of very coarsely crystalline igneous rock, called *pegmatite,* are seen in the canyon walls of the Front Range, such as Bear, Clear and Boulder creek canyons. They were injected as molten rock into cracks in the gneiss and schist. Some pegmatites contain gemstones, such as topaz, garnet, amethyst and amazonstone, while others have been mined for feldspar and mica. Great masses of granite intrude into the metamorphic rocks of the Front Range. All these granites exceed a billion years in age. Most typical are the light gray Silver Plume granite, west of Georgetown; the gray Boulder Creek granite, west of Boulder; and that near Mount Evans. Similar but younger granite bodies, less than 70-million years old, are the source of gold-bearing veins discovered at Central City and Idaho Springs.

The Mineral Belt and vicinity is a geologist's paradise, and studies of rocks have taught us much about the structure, composition and history of this vast area. In fact, the success of geologists in reconstructing the Earth's story by piecing together information from rocks is one of the wonders of science. The Earth is estimated by geologists to be about 4½-billion years old; the recorded history in the rocks of The Mineral Belt area began less than 2-billion years ago when intense heat and pressure — generated by the forces of mountain-building — contorted and recrystallized the rocks of the Earth's crust. Great masses of molten granite intruded into the upper part of the crust and contributed to the deformation. While this uplift of the area was still in progress, the mountains were, in turn, being eroded away by water, ice and wind. Ultimately, the mountains were reduced to a flat, featureless plain, which was submerged at the beginning of the Paleozoic Era, some 600-million years ago. At this time, a broad seaway engulfed the Mineral Belt and surrounding area. Sediments which became limestone, sandstone, and shale accumulated on the floor of this sea, only to be uplifted and eroded again. After most of these marine Paleozoic rocks had been removed during this second wearing-down process — approximately 300-million years ago — sand and gravel, eroded mainly from the Precambrian core rocks of the planed-off ancient mountains, were deposited by streams. These sediments, now hardened, are called the Fountain Formation. Two major advances and withdrawals of the sea marked the next 200-million years. As the sea re-advanced across the region, during the Permian Period, on-shore dunes and beach sands were deposited. These now make up the Lyons Sandstone. As the waters deepened, muds and limy sediments accumulated on the sea floor, and these became the shale and limestone of the lower part of the Lykins Formation.

In the later part of the Paleozoic Era, the sea withdrew to the east, leaving a lake-dotted coastal plain drained by sluggish streams. In the Triassic Period, the area was a dry land, on which silt and sand of the upper part of the Lykins Formation were deposited by wind and intermittent streams. No deposits of early Jurassic age were preserved; however, in late Jurassic time, the area became a swampy lowland where sediments were deposited, which became the Ralston Creek and Morrison formations. Dinosaurs flourished during the Morrison time, and their fossilized bones are found in the Morrison Formation throughout the Rocky Mountain Region. In fact, complete skeletons have been uncovered.

During the Cretaceous Period — beginning about 135-million years ago — the sea again moved westward across the area of The Mineral Belt, and the ripple-marked sands of Dakota sandstone accumulated near the advancing shore. Dinosaurs, which trod these intertidal beaches, were driven westward by the advancing waters. With the shoreline far to the west, water covered most of The Mineral Belt area, and thousands of feet of sediment accumulated on the ocean floor. These became the Benton shale, the Niobrara Formation and the Pierre shale. They contain abundant remains of marine life, such as the shells of mollusks, and the skeletons and teeth of great fish and marine reptiles. Late in the Cretaceous Period, the entire region began to rise slowly, forcing the sea to withdraw for the last time. As the shoreline retreated eastward across the area, beach sands of Fox Hills sandstone were deposited. Streams then deposited mud and sand on the newly emerged coastal plain. This plain supported a lush growth of vegetation, which was buried and became coal. These are the rocks of the Laramie Formation. As the mountains rose, vigorous streams stripped away the soft sedimentary rocks, and, for the first time since the Fountain Formation was deposited, cut into the hard Precambrian rocks. The eroded material was carried eastward beyond the mountain front, where some of it is preserved in the younger formations of the plains, such as the Arapahoe Formation. After a slow start, the rate of uplift increased during early

DAVID S. DIGERNESS PHOTO

THE NATURAL AMPHITHEATRE of Red Rocks Park—west of Denver—has been the site of many musical concerts in recent years. These beautiful sandstone formations have attracted numerous rock-climbers, some of whom had to be rescued from precarious positions on rock ledges.

Tertiary time, and, about 70-million years ago, the sedimentary layers adjacent to the mountains were thrust by pressures sharply upward into hogbacks or sharp ridges of uptilted sedimentary layers. Molten lava found its way to the surface through vents and great fractures along the flanks of the mountains. Volcanoes, somewhere to the west, contributed to the stream-laid volcanic debris, as well as to layers of volcanic ash and mud deposits of the Denver Formation. Fossilized leaves and twigs of trees that grew in the fertile volcanic soil are preserved in the rocks of the Denver Formation. Just east of the town of Golden are the North Table and South Table mountains, two flat-topped mesas capped by lava flows. Lavas that poured across what was then the land surface hardened and protected the soft underlying area from being eroded away, forming the mesa topography. The river presently known as Clear Creek cut its channel gradually down through the cap-rock separating it and forming the two landmark features. The lava appears to have poured out from great fissures adjacent to Ralston Reservoir north of Golden. These fissures now contain masses of chilled lava called *dikes.* Molten granite from deep within the Earth also was intruded along fractures in the ancient Precambrian core of the range. Hot solutions and gases from these melts (chemical reactions) spread into fractures deep within the rocks, and deposited the rich gold, silver and other minerals in the veins of The Mineral Belt's mining districts.

The shape of the Front Range as we see it today is the result of processes of erosion acting over many millions of years. In most places, the mountains are dissected into peaks, ridges and canyons. Remnants of a gently rolling upland surface, however, are preserved locally in the highest part of the Front Range on ridge tops. This upland surface is a remnant of a former wide-spread erosional plain. Travelers over Rollins Pass have the opportunity to see an example of the rolling uplands preserved along the Continental Divide, and many other passes provide like examples. During the Pleistocene

DAVID S. DIGERNESS PHOTO

THIS SAMPLE from Jesse S. Randall's mineral collection is a high-grade lead-silver ore, containing galena, sphalerite and pyrite in silicic vein material. The gold color at the bottom is pyrite, also known as "fool's gold." This specimen was obtained from the Leadville Mining District.

Epoch — popularly called the Great Ice Age — which continued throughout most of the past million years or more, great masses of snow and ice accumulated along the crestline of the mountains and moved down the valleys as glaciers. These glaciated valleys head in great semi-circular amphitheaters called *cirques.* These spectacular and precipitous features of the high-country scenery can be seen near Mount Evans. As the glaciers moved down the valleys, they acted as giant rasps, gouging and scraping the irregularities from the rocks, and rounding and smoothing the canyon walls to produce broad U-shaped valleys like that of Clear Creek, near Georgetown. This U-shaped valley profile contrasts sharply with the V-shaped profile farther downstream, between Idaho Springs and Golden, where the stream valley was untouched by the glacier.

Heaps of debris strewn along the edges of the ice — left after the melting of the glacier — are called *lateral moraines.* The farthest advance of a glacier commonly is marked by a ridge of rock debris known as a *terminal moraine.* Many moraines effectively dam streams and impound lakes. Echo Lake on Chicago Creek (of the Clear Creek drainage), Brainard Lake (of the South Saint Vrain Creek drainage) and many other lakes in the mountains owe their presence to such natural dams.

After the retreat of the valley glaciers, some 6,500 years ago, the streams continued their work of down-cutting in the mountains and foothills, and transporting and depositing sand and gravel or sediments throughout the plains.

A record in the rock of geologic history, as previously mentioned, is exposed to view in The Mineral Belt and east of it to Denver — Colorado's capital city. Denver is situated on the High Plains near the east front of the Rocky Mountains. As one travels westward from Denver toward the mountains, successively older rocks are crossed, from the geologically young rocks of the High Plains and the South Platte River valley to the older rocks of the foothills and the ancient rocks of the mountains. Thus, within a few miles, the journey turns back the pages of time in a lifesized textbook that vividly illustrates the geologic events that shaped the landscape.

The Front Range of the Rocky Mountains in Colorado was formed as a result of a vertical uplift. Most of the crustal movement in the Front Range during the last half-billion years or more has been vertical, both up and down. With each uplift of the land, surface forces of erosion attacked the mountains with renewed vigor, and the eroded material was transported to the flanks of the mountains and redeposited. With each sinking of the surface, which permitted seas to encroach upon the area, sediments were laid like a new carpet

DAVID S. DIGERNESS COLLECTION

GOLD PANNING in California Gulch was first done by a party of Gilpin County gold prospectors, who found the gulch to be rich in gold beyond expectation.

on the older rocks of the ocean floor. When the mountain crests wore down, the sediment from them became buried and preserved in the low areas along the mountain front, where one can see them today and interpret their history. Much of the history of these uplifted mountains is written in the sedimentary layers along their fronts. Denver lies in the broad, gentle valley of the South Platte River, about 15 miles east of the mountain front. The river has eroded its valley into the rocks of the High Plains, which slope eastward from the foot of the mountains. Gravel-capped terraces along the South Platte River are remnants of previous valley floors, preserved from times when the river was flowing at various higher levels. As the river slowly lowered its bed, parts of its flood plain were left high and dry. The best preserved of these terraces is called the Broadway Terrace, named after one of the main thoroughfares in the city of Denver.

Downtown Denver is built on Broadway Terrace, and excavations for high-rise office buildings have exposed loose sand and gravel deposited long ago by the South Platte River. These buildings are anchored in the bedrock that underlies the gravel. Fossil bones of mammoths and some split and shaped bone-fragments that probably were tools of ancient man, have been recovered from excavations in the Broadway Terrace, and scrapers, arrowheads and other stone tools of prehistoric man have been found on the surface of the terrace. West of the South Platte River, several gravel-capped surfaces stand 100 to 450 feet above present stream levels and rise gradually westward toward the mountain front. These surfaces are called *pediments*. They were formed thousands of years ago by streams that emerged from the mountains, planed off the bedrock, and deposited sand and gravel on the eroded surfaces.

Rock layers beneath Green Mountain and the Table mountains are nearly horizontal. Immediately to the west, however — as previously mentioned — they bend abruptly upward so that successively older rocks are exposed westward toward the mountain front. The harder layers of rock form sharp ridges called hogbacks; the softer, more easily eroded layers form the intervening valleys. The most prominent hogback is formed of the Dakota sandstone, a beach deposit of an ancient sea. Abundant ripplemarks, still preserved, formed in the shallow water, and footprints of dinosaurs that walked the sandy beaches can still be seen on the sandstone layers. Older than the Dakota sandstone and farther to the west, located north-northwest of Morrison, upturned against the ancient rocks of the mountains, are the colorful red sandstone monuments of the Fountain Formation. These frame Red Rocks Amphitheater, for which the amphitheater is named.

Geologists classify rocks into three great groups according to the major earth processes that formed them. They are called *Igneous, Sedimentary* and *Metamorphic.*

Igneous rocks are formed from molten-rock material that has cooled and solidified. Molten-rock material originates deep within the Earth and ascends, still molten, to lesser depths or sometimes in volcanic eruptions to the Earth's surface. When it cools slowly, usually at depths of thousands of feet, crystals separate from the molten liquid, forming a coarse-grained rock. When it cools rapidly, usually at or near the Earth's surface, the crystals are extremely small, and a fine-grained rock results. Separate bodies of molten-rock material have or acquire unlike chemical compositions and solidify into different kinds of Igneous rock. Thus, because of different cooling rates and differing chemical composition, a wide variety of Igneous rocks were formed. Dissimilar as they are, obsidian, granite, basalt and andesite porphyry are all Igneous rocks. Several hypotheses have been proposed by geologists to explain the origin of the mineralizing solutions. According to one hypothesis, solutions remaining or expelled from molten rock, or *magma*, as it cooled and crystallized many thousands of feet below the Earth's surface, formed mineral deposits. These more volatile parts of the molten rock contain heavy precious metals such as lead, zinc and gold, and are among the last to crystallize. Where fractures formed by rock movements existed in and above the cooling magma, the more mineralizing volatile fractions escaped to higher levels in the crust. The solution, as expelled from crystallizing magma, is chiefly water, but carries with it silica and some metallic constituents which contain precious metals.

Sedimentary rocks are formed essentially at the surface of the Earth, either in water or on land. At no time during their formation are temperatures or pressures especially high, and their mineral constitutions and physical appearances reflect this fact. Sedimentary rocks are layered accumulations of sediments, the sediments consisting of fine-to-coarse fragments of rocks or minerals, precipitated chemical matter, or animal or plant material. Ordinarily, sedimentary rocks become cemented together by minerals and chemicals, or are held together by electrical attraction, but some remain loose and unconsolidated. Layering in sedimentary rocks is normally parallel or nearly parallel to the Earth's surface; if layers are at high angles to the surface, or are twisted or broken, some kind of rock movement has occured since they were deposited. It is easier for one to visualize the formation of sedimentary rocks than of Igneous or Metamorphic rocks because parts of the process occur around us all the time. Sand and gravel layers of beaches or in river bars resemble sandstone and conglomerate. Mud in many mud flats needs only to be compacted and dried to become shale. In a second hypothesis, geologists surmise that when sedimentary rocks are deeply buried and subjected to heat and pressure, some of the water driven off was an integral part of the rock-forming

Lith. by J. Bien, N. Y. Entered according to Act of Congress, in the year 1866, by A. E. Mathews, in the Clerk's Office of the District Court of the United States for the Southern District of New York. A. E. Mathews, Del.

PIKE'S PEAK AND COLORADO CITY.

A. E. MATHEWS SKETCH
DAVID S. DIGERNESS COLLECTION

PIKES PEAK was the first feature that could be seen by the pioneers as they crossed the prairie, "bound for gold and glory" in the Pikes Peak Region. On a clear day, the snow-crowned summit of this awesome pinnacle could be seen from 100 miles away. A. E. Mathews' sketch of about 1860 shows Colorado City, Colorado's first territorial capital, at the foot of the great mountain.

minerals. These hot solutions also traveled upward along fractures and deposit vein material.

Commonly the principal constituents of a fracture filling or *lode* are quartz, or quartz and calcite, but the lode material generally contains some copper, iron, lead and other metal-bearing compounds. In many lodes, or veins, iron-sulfide-pyrite is abundant; pyrite and iron-stained mica have sometimes been mistaken for gold and have been commonly referred to as *fool's gold.* Sometimes, Sedimentary and Igneous rocks are subjected to pressures so intense, or to heat so high, that they are completely changed and they become metamorphic rocks.

Metamorphic rocks originate while deeply buried within the Earth's crust, usually during the long, continued, gigantic series of events which produce mountain systems. The process of metamorphism does not melt the rocks, but instead transforms them into denser, more compact, foliated rocks. (The term *foliated* means the parallel arrangement of certain mineral grains, which give a laminated appearance to the rock.) In metamorphism, minerals are created by rearrangement of mineral components or reactions with fluids that enter the rocks. Some kinds of Metamorphic rock — granite gneiss and biotite schist are two examples — are strongly banded or foliated. Other kinds, such as some hornfels and quartzite, are massive. Under such conditions of stress or temperature, even earlier-formed Metamorphic rocks are commonly changed to new types. A peculiarity of Metamorphic rock is that with increasing metamorphism, a related but unlike series of rocks is formed. Thus, in favorable localities, one can trace a formation from shale through slate, phyllite and biotite-muscovie schist, to biotite gneiss, and know that all the rock types evolved from the same shale. Elsewhere, however, it may be impossible to tell which of two such dissimilar rocks as basalt and limy shale was the parent rock of hornblende schist. These rock-forming and rock-destroying processes have been active for several billion years.

Thus, over the past 1½-billion years, the combined forces of mountain building, volcanism and erosion, and deposition by streams and oceans, winds, frost, ice and glaciers have sculptured our present-day landscape. The results are here in The Mineral Belt for all to see; it is a geologist's and miner's paradise, an area rich in mineral wealth, a spectacularly beautiful scenic wonderland. Today, the geologic history of the region can be read by anyone interested in its landscape.

ALEX MARTIN PHOTO
DAVID S. DIGERNESS COLLECTION

JESSE SUMMERS RANDALL — born April 23, 1848; died January 24, 1939 — was the founder and publisher of the "Georgetown Courier." Over the years, this newspaper carried voluminous reports about the mining and railway activities of The Mineral Belt, and much of this material has been included in this book in one way or another.

Chapter II

The Arrival of Jesse Summers Randall

NO BETTER PLACE THAN HERE, in the first volume of *The Mineral Belt*, does the following data fit. In Jesse Randall's prized library of personal books was found an intriguing medium-sized leather-bound book of 307 pages, somewhat aged, containing a remarkably dedicated work. Printed on high-quality paper and using the finest printing techniques of 1879, it covers the complete geneology of a branch of the Randall family. This book was a product of many, and is under special obligations to Mr. John Ward Dean, librarian of The New England Historic Genealogical Society, for access to books and documents in his charge, and to Mr. Frank Eldredge Randall, of Boston, Massachusetts. Also included is an up-date by J. S. Randall and his daughter, Pearl.

This geneology has the Randall name going far back in English history. John Randall, the first member of the Randall family to come to America, arrived in the 17th Century. He was a London businessman who had been burned out in the great London fire which burned from the second to the sixth of September in 1666. This was one of the world's worst fires, destroying 89 churches, 13,200 homes and many private and public buildings. As many as 400 streets were laid waste — the ruins covered 436 acres — and 200,000 persons were rendered homeless. From John Randall, who came to the colonies to begin life anew, and who was one of the first settlers of Westerly, Rhode Island, Jesse is descended.

Jesse's father, Abram Joshua Randall, was born at Westmoreland, Oneida County, New York, on March 31, 1823. After ill health prevented graduation from Hamilton College, he went to Kentucky and taught in country schools in Fleming and Bath counties. While teaching in Fleming County, he married the daughter of the aristocratic plantation family of Summers. Jesse, the first of six children born to Mary Ann Summers and Abram Randall, was born on the plantation on April 23, 1848. In 1854, the family returned to New York, where Abram taught school near Rome for three years. In 1860, they moved to Iowa, where he taught school in Bradford County and in Mountair, Iowa. In 1862, when Jesse was 14 years old, Reverend B. C. Galliday was attracted to him and induced the father and mother to let Jesse accompany him and learn the printing business. The boy's first job was on Galliday's *American Union*, at Sidney, Iowa. He later accompanied his parents to Osceola and worked on the *Sentinel* of that town until 1865, when he went to Des Moines, where he was employed by the Des Moines *Tribune*, and remained there for several years. Possibly he would have stayed in Des Moines had not the lure of the West attracted him — an attraction which was strengthened by the fact that his father was there.

His father earlier had heard of Georgetown, in Colorado Territory, from friends who had been there at the time of the great discovery of silver. They told him of an opening there for a school teacher. On June 29, 1867, Miss L. H. Lander, the first teacher in Georgetown, slipped on a log footbridge and fell into Clear Creek, which ran through the center of town. It was running nearly bank-full at the time and the teacher was drowned in the swift current. For reasons of health, A. J. Randall was told by doctors to go to a higher, curative, invigorating, dry mountain climate. Abram Randall took his doctors' advice, and Georgetown seemed to fill the bill. It was a place where he could put his teaching abilities to work and make a living. He became the unfortunate woman's successor.

And so, in 1869, Jesse Summers Randall headed for Georgetown, where his father was teaching as the town's second school teacher. In preparing for his trip, Jesse found that he could go by an easier method than by covered wagon, the way his father had used. He learned that the Union Pacific Railroad, on the 10th day of May in 1869, had finished tracklaying — completing the first great U. S. transcontinental rail line, 1,927 miles in length, the main divisions known as the Union Pacific and Central Pacific. The mainline of the Union Pacific Company extended from Council Bluffs, on the Missouri River, to near Ogden, in Utah Territory, a distance of 1,032 miles. The construction of the Union Pacific began late in 1865, and for some years afterward, there occurred some of the grandest exploits in the way of rapid railway building that the world has ever witnessed. The tracklayers from the east and west met on the northern border of the Great Salt Lake.

In Jesse's diary, he mentions that he traveled west with Charley Cuffel, a printer, on the new railroad and noted that the Union Pacific followed the Platte River Valley some 400 miles west of Omaha. It left that stream at Julesburg, located

DAVID S. DIGERNESS COLLECTION

THIS PORTRAIT of Abram Joshua Randall was taken when he was the schoolmaster at Georgetown. He was born at Westmoreland, New York, on March 31, 1823, and died at Georgetown on January 6, 1925, at the age of 101, a true Colorado pioneer.

just within the limits of Colorado, and then headed up over the great inland plateau to the railroad town of Cheyenne. In his diary, he marveled at the beautiful train, and noted that it stopped to wait for buffalo crossing the track, and that he saw some of the animals shot right alongside the train from the window of a coach. Arriving later at Cheyenne, he mentioned that it was a tent-town, mixed with new buildings, and he made a special effort to get a better look at the locomotives before leaving by stagecoach for Denver City. He had no idea his interest in this railroad would later be renewed and that he would report on the progress of all the new railroads which would be constructed in Colorado. The fare to Georgetown — his destination — was $32.00 by Wells Fargo & Company stagecoach. On the horse-drawn coach a congenial group accompanied Jesse and his friend, Cuffel, and it seems that the Denver, or Cherry Creek, flood of 1864 was still a topic of conversation during the day-long trip and again in Denver on Jesse's night stop-over. Being a newspaperman, he was interested to find out that the offices and facilities of the *Rocky Mountain News* — the territory's first newspaper — were swept away in that devastating flood.

Back in 1860, Denver City was in a wild country, without even a rudimentary territorial government. Although the gold discoveries of 1858 had occurred in extreme western Kansas Territory, no county organization existed. The whole area was simply called the Pikes Peak Region —— hence the saying, "Pikes Peak or Bust," which the '59ers used as a slogan in coming west to seek their fortunes.

Jesse's friend, Cuffel, remained to work for the *Rocky Mountain News*, which had started again after the 1864 flood, and Jesse boarded the stagecoach bound for Georgetown. The stage was a handsome Concord coach, filled with passengers and hauled by a team of four well-matched horses. Before the stage trip was over, however, it proved to be tedious and somewhat disagreeable at times, because of dust kicked up by the horses. But it was one of Jesse's most memorable trips, over steep and difficult roads, and seemingly impassable hills and defiles.

In Jesse's diary he later remarked that ". . .we came by way of Golden, Mount Vernon Canyon, past the town of Mount Vernon, which was the first residence of the first governor of Jefferson Territory, Robert W. Steele, afterward a resident of Georgetown and one of my good friends; through Bergen Park, down Floyd Hill to Clear Creek, passing Spanish Bar, where we were told that the first murder in Clear Creek County was committed; and went through Idaho Springs, which in June of 1869, was a very small village. It took the greater part of a day for the trip of less than sixty miles of winding toll roads. We had lunch at Fall River House where

DAVID S. DIGERNESS COLLECTION

THIS TINTYPE of Jesse S. Randall was taken at the time he started in the printing business at Galliday's "American Union" newspaper in Sidney, Iowa. He moved up to the "Des Moines Tribune" in 1865, and then, in 1869, he moved west to Georgetown, Colorado Territory, to work on the staff of "The Colorado Miner," a pioneer newspaper of the region.

W. H. JACKSON PHOTO
DAVID S. DIGERNESS COLLECTION

END-OF-TRACK on the Union Pacific mainline during construction was always a scene of great activity. This photograph, taken circa 1868, shows a supply train at the railhead in southern Wyoming. Freight wagons carried trainloads of supplies from the end-of-track to the grading-construction camps farther west.

the horses were changed. Everything was new, all the way out; habitations were mostly of log cabins but with some fine buildings here and there such as the white two-story frame stage station hotel and eating house called Fall River House located at the toll gate of the next section of road at the junction of Fall River with Clear Creek, the only spot in the United States where a river drains into a creek. This was also a small thriving mining camp and the gateway to the upper Clear Creek County region.

"A fact about Fall River should be pointed out here: Clear Creek, as mentioned previously, was earlier known as Vasquez Fork of the South Platte River. Fall River emptied into it. When Vasquez Fork was renamed and lost its distinction as a river, Fall River continued to be known by its original name and so the reason why a 'river' enters into a 'creek.' Along our way again we held up at the Downiville stage stop where one of our wheels was repaired."

Arriving in Georgetown late in the day — in a cloud of dust — Jesse was greeted to the little mountain town by his father. Here was the country that was to become the center of Jesse's future life. Lode mines in Clear Creek County had just started, so mineral production was small. Jesse wrote: "Much to my surprise there were a great many women in Georgetown, and the population was right around 350."

The majority of the pioneers came to Georgetown for reasons other than those that brought Jesse to this settlement, and the same was true for Jesse's ailing father, Abram, who came with the intention of teaching in Georgetown's first schoolhouse, as well as for the healthful climate that Colorado had to offer him. A. J. Randall was a scholarly man with a strong bent for geology and mineralogy. In fact, he spent much of his lifetime tramping the mountains in search of specimens and fossils. One of his attainments was a long lyrical poem in the Hiawatha tradition, which has been pronounced first-rate literature. At the age of 90, he walked the steep trail to Green Lake, and if you think that is not a chore for a vigorous man, try walking it sometime. A month or so after Jesse's arrival, the statistics on the schoolhouse in which A. J. Randall was teaching were recorded as follows: "July 3, 1869: Second Ward School; A. J. Randall, Teacher;

DAVID S. DIGERNESS COLLECTION

A HANDSOME CONCORD COACH and a well-matched team of horses was the typical mode of cross-country travel for many years — even after the initial arrival of the "steam cars." Mail, express and baggage, as well as all varieties of people used this bone-shaking conveyance. Jesse Randall boarded a coach similar to this one during his first trip to Georgetown in 1869. It is easy to see why railway lines were so anxiously promoted by the people — and by the press — of the day.

Total Scholars — 52; and Average Attendance — 36." (The school, built of logs, had been opened in 1867.) Abram Joshua Randall also was a gardener at heart and grew some of the first domestic flowers in the area, and, in later years, had his own greenhouse.

It is interesting to note that the high, dry climate surely was curative for Abram Joshua Randall. When he died in Georgetown on January 6, 1925, he was within three months of reaching his 102nd birthday. However, the Randalls are long-lived. Abram's mother, who was born in 1799 and died in 1900, also lived to be over 101 years old.

Jesse's Reminiscence

Another brief item in Jesse's diary, a number of years later, describes his arrival in Georgetown and falls right into this spot:

Father was boarding in the lower part of town with the family of George Reynolds. Like many other miners here, Reynolds was a Cornishman. I boarded with them until Father and I went to batching in a log cabin which stood opposite to the home where I now live.

Father had thirty to forty pupils, one of whom, was a little Elsie James, her father, William H. James, was superintendent of the Terrible mine; he had come over from Central City. He was afterward one of the founders of the Denver National Bank, having made a fortune in Leadville. Elsie was the darling of the town and created a lot of amused laughter while teamsters swore. When the burros, "jacks" we called them, came down loaded with ore from the mines, before they could be gathered after unloading for the return trip, she would chase them and ride them through the streets, they kicking, bucking and braying, her hair flying and her yells of delight rising above all other noise.

Father had arranged for me to take charge of the printing department of *The Colorado Miner*, which I did at once. It had been established in the lower part of town, named Georgetown after George Griffith who, with his brother, D. T. Griffith in 1859 had found here the auriferous veins that eventually led to the discovery of the silver-bearing lodes that gave the world one of its richest silver districts. The upper part of the settlement was named Elizabethtown, for the wife of George, and Griffith Mountain in honor of both brothers. The two villages were consolidated in 1868 as Georgetown.

Governor John Evans of Colorado was financially interested in Georgetown's first newspaper. F. R. Waggoner and Dr. J. E. Wharton, the latter having been employed on *The Rocky Mountain News* in Denver, started *The Colorado Miner* which made its appearance in May of 1867 on the 30th day. The two did all the work in a log building little larger than a shed. The equipment, hand press of the Washington make, and second-hand type, representing in all a worth of $400.00 or a little more. In June, Wharton bought Waggoner's interest and took as a partner A. W. Barnard who had worked on *The News* as a typesetter and who had earlier made a gold discovery near Downiville, the stage stop located down the canyon from Empire Forks of Clear Creek where he had built a small smelter in 1866. Barnard's wife, son and daughter were living in Denver.

In August of 1868, Wharton disposed of his interest and A. W. Barnard and Company became publishers. Barnard borrowing his capital from Governor Evans who was the "company." Evans' interest later passed to William Spruance, Clear Creek County's first recorder. Spruance put in his nephew, Alexander Spruance, in May of 1869. Wharton, after leaving here, attempted to establish a temperance paper, *The Pledge*, at Boulder, but it passed out, and he subsequently became editor of *The Boulder Camera*, and eventually went to Prescott, Arizona. I saw him in the latter seventies when he returned to Georgetown for a visit. I don't know what became of Waggoner."

DAVID S. DIGERNESS COLLECTION

JESSE RANDALL'S travels radiated from Georgetown — known as the "Silver Queen of the Rockies" — in the heart of The Mineral Belt. This was Georgetown in the late fall of 1869, the third-largest town in Colorado Territory, outranked only by Denver City and Central City. New houses were being constructed in anticipation of the long winter ahead.

In June, 1869, Morris E. Ward, a nephew of Governor Evans, succeeded Spruance. Jesse joined the staff of *The Colorado Miner* then being run by Barnard and Ward, editors and proprietors. On June 30th, a daily evening paper was launched, but discontinued for lack of patronage. It returned to a weekly the following October.

Jesse Randall came with the idea of looking out for his father and put his talents to work at Georgetown's first newspaper. Jesse took charge of the printing department of *The Colorado Miner* as foreman-compositor in 1869 and worked on that paper until 1875 when he left it to open a job office. In a few years, after many interviews with the early pioneers who came before him, he became well known as an authority on the history of the area. He traveled much in this region and seemed to prefer the Denver area, where men congregated for business or pleasure from mining camps in all quarters of the territory. Two years in the job-printing business gave him funds and opportunity to establish a newspaper, *The Georgetown Courier*, the first number of which came out May 24, 1877. From that point, his newspaper was closely identified with the remarkable history of that pioneer community. It soon became one of the best weeklies in the state, with an extensive circulation. A little Washington hand press, along with a few cases of second-hand type which originally belonged to *The Rocky Mountain News* of Denver, turned out the first number of that remarkable paper. It had been freighted by wagon to Georgetown in 1867 and stood in a small shed, where *The Colorado Miner* was first printed. Jesse bought the original "plant" of the *Miner* after it had been discarded for new equipment. Until then, the *Miner* had been the only newspaper of the great silver camp.

The Georgetown Courier was a crusader for decency. As a result, Jesse made numerous powerful enemies. The fight against him extended to Washington, D. C. He became known as the "fighting editor." Randall was a close friend of U. S. Senators Edward Wolcott (with whom he had worked on *The*

W. G. CHAMBERLAIN PHOTO
DAVID S. DIGERNESS COLLECTION

THE ROCKY MOUNTAIN NEWS located their business in the Wooton Building, situated in the bed of Cherry Creek prior to the great flood of 1864. Years after this flood wiped out the "News" offices, when a Colonel James Archer was sinking a well at the foot of Fifteenth Street for the first waterworks, his workmen struck part of what had been the old hand press belonging to the "News" — some 12 feet below the surface. About the same time, parts of the newspaper's large cylinder-press were found in a sandbar in the South Platte River, opposite the lower end of Twentieth Street.

Colorado Miner) and Henry N. Teller, and nearly every famous character of the Colorado mining camps. He was the only man with whom Louis Dupuy, the French nobleman, would entrust the secret of his identity. Louie Dupuy built the Hotel de Paris of Georgetown, which became famous throughout the United States in the 1870's and '80's.

Another of his friends was President U. S. Grant, who spent several hours with him on one of his visits to Georgetown. Two other famous people of Jesse's acquaintance were Governor John Evans and Horace A. W. Tabor.

During his lifetime, Jesse collected much material connected with the history of his era, of which a great part was donated to the Denver Public Library, the State Historical Society of Colorado and the Library of the State Supreme Court. One item of particular interest given to the State Historical Society was a blood-stained note pinned to the breast of a Georgetown criminal before he was killed by vigilantes — self-appointed policemen of the area. Other articles of historical significance, which were given away before his death, included some of his scrapbooks and books on local history, duplicate newspaper files, and letters from many of Colorado's celebrated personages. But, being sentimental, he held back some of his choice items for his daughter, the late Mrs. Pearl Sidney of Georgetown, and his granddaughter, Helen Sidney Digerness, originally of Georgetown, who now lives in Denver.

Jesse was a staunch, militant Republican throughout his life, and with but two exceptions, supported the party. One of these exceptions was his firm belief in the theory of "free silver." Throughout the years, and especially in the 1890's, he campaigned vigorously with and for the free silver advocates; but, much to his disappointment, the plan never gained widespread popular support outside the Rocky Mountain West. Jesse was well acquainted with the famous populist, William Jennings Bryan, who spoke on two occasions before Georgetown audiences on free silver. It might be noted that Colorado overwhelmingly went over to the Populist Party in 1892, electing Davis H. Waite governor on a "free silver" platform.

On one of Jesse's early trips to Denver — always interested in other area newspapers — he learned that John L. Merrick was the first to have set up a printing press in the territory, at Cherry Creek. Mr. W. N. Byers and Mr. Thomas Gibson arrived shortly thereafter with their larger equipment and more abundant resources. Comparatively few people know that the proprietors of *The Rocky Mountain News* were not the first to set up a printing press in the pioneer community at the mouth of Cherry Creek; that honor belongs to John L. Merrick, who published the first and only issue of the Cherry Creek *Pioneer* on the same day that the first number of the *News* was given to the public. The first copy of the *Pioneer* was actually off the press nearly half an hour before the *News* made its appearance, though this claim was challenged by the *News*.

A newspaper in the mountains named *The Rocky Mountain Gold Reporter and Mountain City Herald* began its short career early in August with the first number dated August 6, 1859, just three and a half months after the initial publication of the Cherry Creek *Pioneer* and *The Rocky Mountain News.* It was located at Mountain City, a pioneer settlement in Gilpin County that was absorbed by the town of Central and became known as Central City. Central City was later well represented in the newspaper world by the Central City *Register*.

The Gold Reporter used the printing plant brought to Colorado for the Cherry Creek *Pioneer*. This plant had an interesting early history; however, after that, no one knows what happened to it. It was used by *The Gold Reporter* until October, 1859. Not long after that, Capt. George West, at Golden, contracted for Gibson's press and supplies for the winter of 1859-'60, and on the 7th of December issued the first number of *The Western Mountaineer*. He continued its publication in an enlarged form until the following fall. The outside world heard the noise of the Pikes Peak Region from newspapers like the *Mountaineer*. The *Mountaineer* and the *Gold Reporter* were both four-column, four-page papers. The columns were 12¾ inches long and 11 picas in width (slightly less than 1-7/8 inches, or 46 mm.), as compared with the standard newspaper column, which was 13 picas in width (about 2-3/16 inches, or 55 mm.). The *Mountaineer* was short lived, its final issue being No. 23, dated November 29, 1860.

The Rocky Mountain News, which recorded the beginnings and ends of many newspaper ventures, announced in its issue of March 11, 1863, the appearance of another newspaper of the same name, *The Western Mountaineer*, published in the Buckskin Joe Mining District by Mat Riddlebarger and L. B. St. John. One issue of this newspaper, an independent Democratic Union paper, was dated at the old town of Laurette in the Buckskin Joe district of South Park. Another paper of the four-column, four-page style, the first issue of which was dated August 1, 1861, was the Colorado City *Journal*. This was a paper with a lot of spirit, and the editor, B. E. Crowell, was well versed on the needs of the new country and expressed his ideas in his editorials with great taste and skill.

Mr. Thomas Gibson, already mentioned as one of the publishers of the first copy of *The Rocky Mountain News*, was a good promoter of newspapers. He also is credited as the founder of *The Gold Reporter* and was the founder and sometime editor of *The Colorado Republican and Rocky Mountain Herald*, the first daily newspaper published in Denver. The *Republican's* first issue was dated May 1, 1860.

The Cherry Creek Flood of 1864

Jesse was still interested in the story he heard about the Cherry Creek flood when riding the stagecoach into the territory. So, on later trips to Denver, he investigated and recorded some very interesting facts. The history goes back to June 6, 1860, when Editor W. N. Byers of *The Rocky Mountain News* proudly announced in his newspaper, "Our office is in the large frame building in the middle of Cherry Creek, under the sign of the American flag. All arrivals are invited to call and register." Mr. Byers was confronted by an old Indian, who called upon him about this time to compliment him on the newspaper. He seemed to understand the paper sufficiently to give it his approval — "heap good medicine" — but could not understand why the white men were building houses in the very creek bed. He solemnly warned of "heap big water," such as he had seen cover the whole bottom lands "so," and he held his hands high above his head. Byers was a strong advocate of consolidation of the then rival settlements of Auraria, on the west side of the creek, and Denver City, on the east side of the creek; and, in token of his absolute neutrality, built his newspaper office on the boundary line.

On August 1st of that year, following a heavy rain on the headwaters of Cherry Creek — to the south of town — the stream flowed with moderately strong current for a time and Byers mentioned that "Cherry Creek appeared to present a rather serious problem, for we had a demonstration of what could be expected from a heavy rainfall on the Divide to the south, though we were not yet inclined to believe the Indian claims that the whole settlement was subject to flood." About this time, the American Civil War, Indian depredations and many new gold strikes in the mines occupied the attention of the public mind to such an extent that Cherry Creek seemed forgotten.

Owners of lots abutting the creek encroached on what was commonly called the channel by building structures that narrowed the natural course of the drain to what was considered a reasonable streambed for the water that ran in it. On the afternoon of May 19, 1864, a moderate rain occurred in Auraria and Denver City. For several hours, heavy black clouds obscured the area, and frequent rumblings of thunder and lightning were seen and heard. In the evening, the saloons, gambling houses and other places of entertainment had their usual crowds. There were no disturbances and the night sky was clear as the moon rose above the settlement. By midnight most of the town's people had retired for the night. Suddenly, those who were still awake heard a strange sound in the south, like the noise of the wind, which increased to a mighty roar as a great wall of water — a flash flood — came thundering down Cherry Creek, bearing on its crest trees and other drifting objects, rushing toward Auraria and Denver City. The wall of water hit the settlement with great force. The building of *The Rocky Mountain News* office and its contents were distributed along the sand bars, banks and in open fields of the South Platte River flood plain, along with other buildings, parts of houses, dead horses, cattle and sheep.

Mr. O. J. Goldrick told the story of the great flood in the *Commonwealth* newspaper, in the issue of May 25, 1864. Up to this time, that paper had been a rival of the *News*.

W. G. CHAMBERLAIN PHOTO
DAVID S. DIGERNESS COLLECTION

DURING THE EARLY MORNING HOURS of May 20, 1864, a great wall of water — a flash flood — came thundering down Cherry Creek into Denver. This viewpoint was at Fourteenth and Lawrence streets, and the photographer pointed the lens of his camera west, over the Auraria district.

A frightful phenomenon sounded in the distance and a shocking calamity presently charged upon us. The few who had not retired to bed, broke from their bulidings to see what was coming. Hark! What and where was this? . . . Oh, it was indescribably and inconceivably awful to behold that spectacle of terrible grandeur, as the moon would occasionally shed her rays in the surges of muddy waves whose angry thundering drowned all other noise. Now the torrent swelled and thickened, sweeping tremendous trees and dwelling houses before it — a mighty rush of impetuous water, wall-like in its advancing front as it rolled with maddened momentum toward the Larimer Street bridge. Next reeled the dear old office of *The Rocky Mountain News*, and down it sank, with its Union flag staff, into the maelstrom of surging waters, soon to appear and disappear between the waves.

Mrs. Sanford, who was located at Camp Weld just south of Denver at the time, wrote in her diary some interesting details:

On the night of the 19th, the watchman at the Government Corrall pounded on our door with the startling intelligence that a great flood was coming down Cherry Creek and many people were drowning. We thought him fooling until the roar of the waters could be heard as of a mighty tempest. Mr. Sanford rushed to the creek but returned quickly, saying I must see the awful sight. We found hundreds of people along the creek banks. Many of the women and children in their night clothes, having been rescued from their homes below by the cavalrymen from the First Colorado Regiment barracks. By the light of bonfires along the creek, we could see the inky waves, fifteen to twenty feet high, carrying trees, houses, cattle and sheep, and for all we knew, human beings, to certain destruction. It was a wild, weird night, never to be forgotten. Early dawn revealed scenes of pitiful desolation, and the area still flooded.

In the confusion of rescuing families, many were divided and plaintive cries of children for their parents, wives for their husbands and little ones, could be heard above the noise of the yet foaming and roaring waters.

Early in the forenoon of the 20th, several unsuccessful attempts were made at improvising a way of crossing the flooded area. Finally, a stout cord was thrown across the creek at its narrowest point. By gradually increasing the size of ropes, which were pulled back and forth, the final two-inch ferry ropes were securely anchored and a rope bridge was swung over the now receding, but still dangerous, waters. Mr. and Mrs. William N. Byers, with their two children, Mollie and Frank, reached town, from where they had been rescued by Colonel Chivington's cavalrymen. His home was located

W. G. CHAMBERLAIN PHOTO
DAVID S. DIGERNESS COLLECTION

THE GREAT CHERRY CREEK FLOOD was still very much in evidence at dawn on May 20, 1864. This picture suggests the character and extent of "West Denver" at this time.

south of town, in the river bottoms on the east side of the Platte, which also flooded. When rescued from their perilous position, the river had cut a new channel and left them on the west side.

Denver City's early-day photographer, W. G. Chamberlain, recorded the great 1864 flood in the superb photographic artistry he was known for. Since 1864, there have been but three instances when high water occasioned very serious damage to Denver along Cherry Creek and the South Platte River: that of May 22, 1878; July 14, 1912; and June 6, 1965. The damage in 1912 amounted to several-hundred-thousand dollars, with some loss of life. Estimates of the peak discharge through Denver at that time ran as high as 30,000 cubic feet per second. The flood of 1965 was the greatest, but affected an area only along the South Platte River. Denver now was protected from Cherry Creek by the large Cherry Creek Dam. It was built by the U. S. Corps of Army Engineers as a flood control reservoir southeast of downtown Denver. The flood waters were trapped before reaching the city, and almost filled the reservoir. When this earth-fill dam was being built, it was looked down upon by many as a waste of money. The '65 flood proved the value of this dam by saving millions of dollars in prevented flood damage and more than paid for the project. Today, a recreation pool has been established in the bottom land behind the flood control dam, in the area which experienced the '65 flood. The recreation area and lake created by the dam provides boating, fishing, swimming, water skiing and picnicking.

The '65 flood has since caused the U. S. Corps of Army Engineers to construct two additional flood control dams to protect Denver. Both dams inundate the old right-of-way of the Old South Park Railroad. The biggest of the now completed projects is Chatfield Dam across the South Platte River, downstream from the confluence of the two Plume Creeks with the Platte. This project included the creation of a new recreational area behind the dam. This dam is named for Isaac W. Chatfield, one of the DSP&P's early railroad contractors who actually constructed the railroad through the area covered by the new reservoir. The other dam has been constructed on Bear Creek and is named Mount Carbon Dam. This dam has just been built east of Morrison and will inundate the old roadbed of the Morrison line along Bear Creek. A recreational pool behind the dam is planned for the near future.

THE COURIER

PUBLISHED THURSDAYS.

OFFICE:
Alpine Street, Opposite McClellan Hall.

TERMS:
One Year $2.50
Six Months 1.50

J. S. RANDALL.

Georgetown Courier.

VOLUME III. GEORGETOWN, COLORADO, THURSDAY, JUNE 12, 1879. NUMBER 4.

MASKED BATTERIES.

If you'll keep it secret, honor bright,
I'll tell you a little story, so:—
Something that happened to me last night,
Here at the masquerade ball, you know.

[illegible]

LUCY'S FOLLOWER.

"Upon my word, Mr. Braintree," said Mrs. Postlethwaite, "that little maid who waits at our table has the prettiest face I ever saw."

"She is pretty, ma'am," acknowledged the husband of the "fuel Block"; "but she ain't a common servant neither. Her uncle owns the mill down at the Corners, and she lives out during summer to get pin-money for ribbons, and all that sort of thing."

"I need a girl," said Mrs. Postlethwaite. "I wonder if I could take her home with me when I go."

"Lucy Eastburn would jump at the chance," said Mr. Braintree. "She always wanted to go to London."

And pretty Lucy colored like a daisy, and smiled radiantly, and said, "Yes, ma'am, if you please," when Mrs. Postlethwaite asked her if she would like to go to town. And then she ran home with all her might to tell Uncle Reuben and Aunt Cydonia.

"Well, I declare," said Cydonia, "you are in luck, Lucy."

"I don't know about that," said Reuben, solemnly. "The city's a big place, and Lucy always was a feather-headed thing. I think you are better off here, my girl."

"Oh, uncle," cried Lucy piteously. "Oh, please, please let me go. I've always so longed to get to the city, and you have believed so—"

"Let her go, Reuben," said Cydonia, who in her secret heart, sympathized with little, trembling, excited Lucy. "Mrs. Postlethwaite's an elegant lady, and there can't no harm come to her. And it's a fine chance, I'm sure, for Lucy to see something of the world."

So Reuben took Lucy down to the village to buy dimity for a store of white aprons; and when Mrs. Postlethwaite went back to town Lucy sat beside her in the train.

"Now, Lucy, this won't do," said cook.

"What won't do, Miss Priscilla?" said Lucy Eastburn, scrubbing diligently.

"This business of sailor fellows leaning over the area railing to talk whenever you're sent out on errands," retorted cook, sourly.

Lucy tossed her pretty, sunshiny head. It was simply intolerable to be lectured by a cross, middle-aged old maid.

"I shall do as I please," said she; "and it's no affair of yours, Priscilla Pope."

"It is of missus, though," said cook, dryly; "and I shall just let her know of all these goings-on."

So pretty Lucy was called up into the boudoir and solemnly "spoken to" by Mrs. Postlethwaite herself.

"You are only a child, Lucy, and you don't know the world," said Mrs. Postlethwaite. "Priscilla is a much better judge of things than you."

"Please, ma'am," said Lucy, "I don't think it is any of Priscilla's business."

"Lucy," said the lady, gravely, "I am sorry to see you take well-meant advice in such a spirit. I wish you to understand that I forbid all followers, and hangers-on in the kitchen from this time henceforward."

And Lucy Eastburn went down stairs with crimson cheeks and heart full of vague resentment.

"Because I'm only a servant they want to keep me down and tread me under their feet," said she to herself; "but I'll show them that I can be as independent as any one."

She was tripping home that very afternoon with a pitcher of cream, when her sailor admirer lounged up to her side—the dark, Spanish-faced fellow, whom she fancied must look exactly like Byron's "Corsair."

"*Lucia mia*," said he, "why do you avoid me thus?"

"I don't avoid you, Antonio," said she. "It's Mrs. Postlethwaite and Priscilla, the cook. They won't let me go to the front of the house if they can help it."

"Ah!" said Antonio Ceccare, with a shrug of the shoulders; "I see. Jealousy."

"Yes," said Lucy.

"But, see here, Lucy, cara," said the dark-faced man, "true love is not to be conquered thus."

"No, of course not," said Lucy.

"Miss Priscilla is going to her cousin's to-night," said Antonio.

"How did you know?" said Lucy, with wide-open eyes.

"And Mr. and Mrs. Postlethwaite have made up a party for the opera."

"It's true enough," said Lucy, thinking that the Spanish sailor must be a conjuror.

"And you'll be alone in the kitchen."

"Yes," said Lucy.

"Very well," said Antonio. "When you hear the whistle of a robin in the area you will know that it is I."

And he went off with an unconcerned air, just as Miss Priscilla popped her head out of the door to see what it was that kept Lucy so long with the cream.

All that afternoon Lucy went about feeling exactly like the heroine of a story.

She had a real lover, and she was to have a secret meeting with him, in spite of jealous Miss Priscilla and her unsympathetic mistress, once in a while.

Conscience would give her a warning prick, and it would occur uneasily to her mind that it was not exactly the wisest thing in the world to set herself directly in opposition to those who were older and wiser than herself, but Lucy was only 18, and her pretty head had been fairly turned by Antonio Ceccare's flattering words.

And so, when she heard the low, soft whistle in the area outside at 9 o'clock that night, she unbarred the door and admitted Antonio at once, but, when she saw that he was not alone, she started a little.

"Who is that man?" said she.

"It's our First Lieutenant," said Antonio, soothingly; "Senor Jose Garcia, who is anxious to be introduced to you."

And Senor Jose came and made himself very much at home with a plate of cakes and a pitcher of beer which Lucy had provided for the occasion, for she was a deal too honest to touch anything out of Mrs. Postlethwaite's store.

But as Senor Garcia was drinking pretty Lucy's very good health the glass slipped from his hand and shivered in pieces on the floor.

"How awkward of me," said the Senor, looking up at a row of glittering tumblers on a closet shelf opposite, "to put la senorita to the trouble of getting me another."

"Oh, it's no trouble at all," said Lucy, coloring and bridling with delight at the idea of being called "la senorita." And she jumped up and ran into the closet.

But just as she was reaching up to the tumbler she found herself in darkness. The closet door had been abruptly closed upon her, and she could hear the turning of the key [illegible]

"It isn't a joke at all, my dear," said Antonio Ceccare. "It's business. Just you keep as quiet as you can while me and Senor Jose goes up stairs and looks through my lady's silver-closet and jewel-cases. If you raise a squeak, there's that in your head you won't be hurt; otherwise I won't answer for the consequences."

But Lucy Eastburn, undismayed by the covert threat, began to shriek and scream at the top of her lungs.

The idea of thieves and burglars in her mistress' house—thieves and burglars, too, who had been attracted thither by her own folly—drove her nearly frantic. But the solid door deadened her cries, and the two men had apparently hurried up the stairs in quest of booty, without pausing to listen to her remonstrances.

Alas, poor Lucy! She could only sob and scream, and strive vainly to shake the solid fastenings of the door in an agony of shame and terror.

Suddenly the key revolved grudgingly, the door was opened, and, to Lucy's intense dismay, a huge policeman stood before her.

"Ah," said he, "you're the maid, ain't you? We shall want you in the witness-box. We've got them fellows safe and tight—followed 'em all the way. Reg'lar thieves dressed up in sailor's clothes, and makin' b'lieve they're Spaniards."

And Lucy Eastburn fainted away and recollected no more.

* * * * * * *

"Going home, Lucy?" said her mistress, kindly. "My poor girl, I have no idea of discharging you. Cook says you have behaved splendidly, and—"

"Yes, ma'am, I know," said Lucy, shuddering all over and turning pink and pale by turns; but, if you please, I don't think the city is any place for me, and I'd rather go back to the country."

And no persuasion or argument could induce Lucy Eastburn to change her mind.

Why Dick Catlin Bathes No More.

It was Sabbath in the little town of Frisco, and a holy quiet settled down as though a mist from heaven, with the sunlight over the Wasatch mountains, wrapped all in peace. It was Sabbath, and out from the little town came the stage—the stage that has rocked and swung and rolled along from the white, billowy beach of the Atlantic, across the prairie farms of the West, up, over and beyond the towering Rockies, and bathed its alkali-stained boot in the placid waves of the Pacific ocean. Dick Catlin drove that stage, and a better driver never whaled the flank of a broncho. The day was warm as up the Beaver river rolled he, and his voice, clear as a trumpet, made the carbonate-stained hills of Lincoln ring again and again as he sung:

Up and down this world I've roamed,
And to many camps I've been;
The winds of many lands have combed
Those shaggy locks, I ween.
I've (get up there, Rowdy!) been to old Montana,
Likewise to Idaho,
But to Frisco give the banner,
Where the big bonanzas grow.
God's blessing on Montana,
God bless old Idaho;
But, hurrah! hurrah! for Frisco,
Where the big bonanzas grow.

The warm sun thawed the old man's heart, and the memory of old times stole into his soul, bringing back again the days gone by, the swimming, apple-stealing, dog-fighting days when he was a boy and Ben. Holladay nowhere. Rounding a point in the river, the shade fell across a nook. Dick pulled up, and, as he had no passengers, concluded to take a swim. He knew his team and they knew him. They had been together for years, and, as Dick dropped the lines, he said to himself: "There is nothin' like knowin' stock that's posted. Stay still now, boys" (this to the horses), "and the old man'll have a good wash." Undressing and piling his clothes on the seat, Dick left the stage and plunged into the cool waters. But the stock "was posted." No sooner had Dick plunged in than they plunged out, and, when he crawled out on the bank, a streak of dust away off through the sage brush was all that the poor fellow could see. Now came the tug of war. The team kept the road, and Dick knew that, by cutting across the country and running (as he afterward said) like blazes, he could head them off. There was no one there to witness the race. Perhaps at times a mischievous angel would peep through his tiny fingers and laugh at that big, burly, stark-[illegible] in the valley of the Beaver. When he overtook the stage and got the team under control, he didn't say much, but, throwing the lap-robe in front of him till a wagon-load of women passed by, he scraped the blood and the cactus and sage and greasewood from his limbs, and swore that, if the good Lord would forgive him, he would never wash any more. And to this day Dick has kept his promise.—*Salt Lake Tribune.*

How to Cook a Steak.

There is no mystery about broiling a beefsteak, and yet ninety-nine times in a hundred it is badly cooked. The simple art is to cook a steak without smoking it, and to retain the juices. When you rest your cake of Indian-ink on a palette, and happen to put your brush in your mouth and taste it, the peculiar savor is that of carbon. Now, carbon is one of the products of the imperfect combustion of hydro-carbon, and the best lampblack is made that way. If there is any taste of Indian-ink about a steak or chop, it is at fault. The art, then, is simply to broil without firing or blazing. No steak can be cooked without watching it. The fire must be very hot. The outside must be well cooked in order to keep the juice inside; but the outside must have scarce any thickness. The thing to do is to turn a steak from side to side, and to keep doing it. Never pepper or salt a steak until it is on the dish. To pepper it might not do so much harm, but to salt it is almost a crime. Never use a gridiron with too broad slats; that scores the beefsteak through, and fries overdone, dry portions. Incline your gridiron. If there is a blaze from the melting fat, the flame will burn beyond the steak. It is difficult to arrive at a proficiency with a steak which has too much fat on the edges. Trim your steak well, and, if there be fat on it, when your gridiron is inclined, let the fat portion be up toward the handle at first. Good steaks must be served instantaneously on a hot plate. Butter is admissible on a steak, but only in very minute quantity. If the crude steak be of good quality, cut an inch and an eighth thick, and, the cooking artistic, the juice of the meat follows every cut. We eat more beefsteaks, butchers know how to cut them more artistically, and we cook them better in the United States than in England. This is in accordance with the latest testimony furnished by Mr. Richard Grant White in the *Atlantic Monthly.*—*Exchange.*

Sheep in Colorado.

A Colorado correspondent of the Troy *Times* gives some timely information on a subject of interest in the trade just now. The most reliable information obtained, he says, gives the number of sheep in the State as about 2,000,000, valued at $2.25 per head, or $4,500,000 in total value. Last year's clip of wool was about 5,000,000 pounds, estimated at 17½ cents per pound, $875,000. There should be added to this 500,000 lambs, valued at $1.50 per head, $750,000. During 1878 30,000 head were driven in from California, and 15,000 from New Mexico. The shipments of wool the coming season will be about 7,000,000 pounds, which at last year's price will amount to $1,225,000, to which may be added the value of 750,000 lambs at $1.50 per head, $1,125,000, making $2,350,000 income from sheep alone. Owing to the long distance and the want of double-decked cars, the railroad freight on sheep to Eastern markets is so high that it will not pay to ship mutton to any great extent.

Thus far the business of sheep-raising in Colorado has been very profitable. A flock of 1,800 ewes, costing $[illegible],500, were placed on a ranch in Southern Colorado. In eight years 1,600 sheep were killed for mutton and consumed on the ranch, and 7,740 were sold for $29,680. There are 14,800 head on hand, worth, at $3 per head, $44,400. The wool-clips paid for shepherds and all current expenses. The result shows a net profit over the original investment of $69,520, equal to 193 per cent. per annum for eight years in succession. Per contra, out of a flock of 1,200 very fine selected ewes, worth $4 per head, 800 died during a storm of two days last March. The 400 that survived raised last summer more than that number of lambs. The dog is a valuable auxiliary in the care of sheep. The "Scotch collie" surpasses all others in his natural aptitude for this work, and oftentimes one well trained sells for $150.

American Agriculture.

The possibilities of American agriculture can best be appreciated by reviewing the facts that the area of arable lands in the United States exceeds 1,500,000,000 square miles, while the area at present under cultivation is estimated to be 174,900,000 acres, or less than 275,000 square miles. It is probable that the next census will show a valuation [illegible] $11,000,000,000, and an annual production of $3,000,000,000. The fact that more then one-half of all our people who are engaged in the ordinary pursuits of life are directly employed in agriculture is at once a sufficient evidence of its universality and importance.

In the year 1878 there were 51,008,000 acres in Indian corn, producing 1,371,000,000 bushels; 32,208,000 in wheat, with a yield of 422,000,000 bushels; 13,176,000 in oats, giving 411,855,000 bushels; 1,790,000 in barley, showing a crop of 42,000,000 bushels; 1,621,000 in rye, with 25,800,000 bushels, and 673,000 acres in buckwheat, yielding a product of [illegible] bushels. Thus the total cereal production for 1878 was 2,284,902,000 bushels, raised on 100,476,000 acres, while in 1874 the total crops were but 1,454,180,200 bushels; in 1876, 1,962,821,600, and in 1877, 2,178,934,646 bushels. The future increase in the farm productions of this country will result largely from improved systems of cultivation, as well as from extended areas under the plow. A system of agriculture presents but little to be proud of which averages to the acre over the whole country a yield of but twelve bushels of wheat, less than twenty-seven bushels of corn, and only twenty-eight bushels of oats. Compared with what modern agriculture is capable of in this direction, it is not too much to affirm that our farm crops, even on their present area, may be doubled in quantity.—*American Cultivator.*

A Female Duel.

A good deal has lately been heard of the progress of female emancipation in Russia, but it is somewhat of a novelty to find the Russian ladies figuring in the character of duelists, as was the case not long since with two belles of Petigorsk, a well-known fashionable resort on the northern slope of the Caucasus. A dispute arose between the rival beauties, springing out of the attentions paid to each in turn by a handsome young cavalry officer quartered in the neighborhood. The quarrel ran so high that one of the Amazons at length dispatched her maid to the other with a formal challenge, which was instantly accepted. The belligerents met without seconds in a lonely place outside the town, each armed with a brace of loaded pistols. Before, however, they had even taken up their respective positions, the trembling of the one lady's hand caused her pistol to explode prematurely, sending a bullet through the dress of the other, who shrieked and fell down in a swoon. The assailant, frightened out of her wits, flung away her weapon and rushed to raise the supposed corpse; but her ungrateful antagonist, recovering her senses as suddenly as she had lost them, clutched her by the hair with one hand, while boxing her ears with the other in the most energetic style. The firing having now ceased, the battle proceeded hand to hand. Locks of hair, ribbons and shreds of clothing flew in every direction, and, but for the timely advent of three or four policemen, the affray might have ended like the somewhat-similar combat of the Kilkenny cats. The military Lothario's only remark on hearing the story was, "It's lucky they took to clawing each other instead of me."

A Big Battle-Picture.

A gigantic picture, representing the battle of Waterloo, is in course of execution at Brussels. The canvas measures 115 metres in circumference and 14½ metres in height, so that its superfices represents 1,667½ square metres. The artists, who are under the direction of M. Castellani, are installed in one of the vast workshops of the Pauwels Company. The painting is done a third at a time. The portion now under the brush is that on which the Belle-Alliance farmstead is shown, with the corpses of Highlanders and French lying thick on the ground, a spot where a desperate struggle took place. Further on we see the battalions of the Guard executing their last charge, and then forming the celebrated square. Napoleon, on his white horse, is in the center of the square, surrounded by his staff.

SABBATH READING.

The Four Calls.

The Spirit came in childhood,
And pleaded, "Let me in;"
But ah! the door was bolted,
By thoughtlessness in sin.
The child said, "I'm too young yet;
There's time enough; to-day
I cannot open." Sadly
The Spirit went his way!

Again he came, and pleaded,
In youth's bright happy hour;
He called, but heard no answer;
For, lured by Satan's power,
The youth lay dreaming idly,
And saying, "Not to-day:
Not till I've tried earth's pleasures."
Again he turned away!

Again he came in mercy,
In manhood's vigorous prime;
But still he found no welcome;
The merchant "had no time,"
No time for true repentance;
No time "to think and pray;"
[illegible]
[illegible] away!

Once more he called and waited,
The man was old and ill;
He scarcely heard the whisper;
His heart was cold and chill.
"Go! leave me: when I need thee,
I'll call for thee," he cried;
Then, sinking on his pillow,
Without a hope he died!

—*National Baptist.*

Of Wisdom and Forethought in Our Actions.

We must not trust every saying or suggestion, but warily and patiently ponder things according to the will of God.

Yet alas; such is our weakness, that we often rather believe and speak evil of others than good.

But perfect men do not easily credit everyone who speaks to them; for they know that human frailty is prone to evil, and very subject to error in words.

It is great wisdom not to be rash in thy doings, nor to stand stiffly in thine own conceits.

As also not to believe everything which thou hearest, nor immediately to relate again to others what thou hast heard or dost believe.

Consult with him that is wise and of sound judgment, and seek to be instructed by one better than thyself, rather than to follow thine own inventions.

A good life maketh a man wise according to God, and giveth him experience in many things.

The more humble a man is in himself, and the more subject unto God, the more wise and peaceful shall he be in all things.—*Thomas a' Kempis.*

The Bright Side.

Look on the bright side. It is the right side. The times may be hard, but it will make them no easier to wear a gloomy and sad countenance. It is the sunshine, and not the cloud, that gives beauty to the flower. There is always before or around us that which should cheer and fill the heart with warmth and gladness. The sky is blue ten times when it is black once. You have troubles it may be. So have others. None are free from them; and perhaps it is as well that none should be.

They give sinew and tone to life, fortitude and courage to man. That would be a dull sea, and the sailor would never acquire skill, where there is nothing to disturb its surface. It is the duty of [illegible] and enjoy [illegible] within and without him; and, above all, he should look on the bright side. What though things do look a little dark? The lane will turn, and the night will end in broad day. In the long run the great balance rights itself. What appears ill becomes well—that which appears wrong, right. Men are not always to hang down their heads or lips, and those who do only show that they are departing from the paths of true common sense and right. There is more virtue in one sunbeam than in a whole hemisphere of clouds and gloom. Therefore we repeat look on the bright side. Cultivate all that is warm and genial—not cold and repulsive, the dark and morose.—*The Interior.*

The Vice of Comparison.

One of the commonest and worst of the vices is the vice of comparison. There is good authority for saying that those who measure themselves by themselves and compare themselves among themselves are not wise. The Pharisee who went up into the temple to pray fell into this snare. "God, I thank thee that I am not as other men are." This is the language of spiritual pride and of censoriousness. Thinking in this strain he is sure to paint others as black as he can in order that he may show himself to the better advantage on this background. It would have been well to thank God for not being suffered to fall into grievous sin; but the comparison between himself and other men was sure to bring with it many harsh judgments and many hateful passions. Such a comparison, moreover, puts the true standard of character out of sight; for that is the law that demands perfection. God's requirement is not, "Be ye good as other men;" it is, "Be ye perfect, even as your Father in heaven is perfect."

Besides, such a comparison is very likely to be unjust. How do you know that you are better than "this publican," Sir Pharisee? Perhaps his life may outwardly seem worse than yours; but it may be that he, though vicious and unprincipled and degraded in the eyes of men, is a better man than you are in the sight of God. It may be that he was born with an organization that inclined him to indulgence. It may be that his environment from his earliest life has been evil. And it may be that against all these hindrances he has striven bravely—far more bravely than you have ever striven; and, though his achievements, so far as man can see, have not been remarkable, God may credit this man, who has lived a bad life, with more real heroism than you have ever shown. You had better not thank God that you are better than other men, for God knows both you and them; and He may know that they are better than you.—*Sunday Afternoon.*

Fashion Fripperies.

No toilet is complete without a jabot of some kind.

Shirred lining for bonnet brims take precedence of all others.

Black toilets and black lace bonnets are the furor of the moment in Paris.

The latest novelties in shoes are the Henri Trois sabots and the Catherine de Medici souliers.

Parisian coiffures are de rigeuer low on the back of the head, or rather on the nape of the neck.

Delicate lace-clocked stockings should always be worn over another pair of tinted silk or Lisle thread.

Novelties in shoes have Louis Quinze heels, and many straps across the instep, with bows and buckles.

Gypsy bonnets have large flowing brims raised in front, tied down at the sides, and falling on the neck, after being partially raised in the back.

Satteens and mummy-cloth novelties show dainty little flower and leaf bouquet and vine patterns in jardiniere colors on pure white grounds.

The favorite flowers for tacking on bonnet crowns are asters, marguerites, pansies and eglantine blooms, with the yellow anthers showing plainly on the delicate pink ground of the petals.

A gay morning robe always looks vulgar at a watering place or hotel breakfast, and corduroy velvet as well as satin is seen in combination with camel's hair. French bonnets and other light wool fabrics in spring costumes.

Dressy parasols for the country or seashore are of ecru pongees, with spots, rings, or diamond patterns of brown, red, or blue, or they are of plain ecru pongees, with bandana or Scotch plaid linings, or gay-colored linings with white polka dots.

Nearly Buried Alive.

The heroine of the following remarkable story—but which comes from unimpeachable authority as perfectly true—says *Chambers' Journal*, was a young lady of 13 or 14 years. After a some[illegible]pearance, died. The mother literally refused to believe it, although the doctors and the other inmates of the house saw no reason to doubt the fact.

The funeral was arranged, the grave made, and the specified three days had come to an end. The mother had never left her daughter's body; she had tried every available means to restore her, but to no avail. As the hour approached for the ceremony to take place, she became more and more distracted, and more desperate in her efforts to convince herself that life still lingered.

As a last resort she went for some strong elixir, and—taking out of her pocket a fruit knife with two blades, one blade of gold, the other of silver—proceeded by continual working to force the gold blade between the teeth; when inserted she poured a drop of the elixir on the blade, then another and another, and tried to make it enter the mouth; but it seemed only to trickle back again and down the chin. Still she persevered, becoming more desperate as the moments flew on to the hour, now so near, when her child was to be taken from her.

At the very last, when she was beginning to fear the worst, she thought she detected a slight spasm in the throat, and on close examination she became aware that the liquid was no longer returning, as it did at first. She continued the application, every moment feeling more excited and more joyfully hopeful. Presently the action of swallowing became more decided. She felt a feeble flutter at the heart, and before long the eyes gradually opened, and closed again; but the breathing became quietly regular, and the mother was satisfied that now no one could dispute the fact; so she called her household round her, and proved to them the joyful fact that her child was restored to her, and that no funeral procession would leave the house that day. Before long the child fully recovered. The fruit knife, with its two blades, is to this day the most precious heirloom in the family possessions.

A Dog Story.

A dog is a good thing to have around the house, and nobody is better convinced of this fact than the proprietor of a certain boarding-house on Grand River street east, not very far from Woodward avenue. It seems that the lady left a [illegible] ago, while temporarily absent from the room, and when she returned the bill had disappeared. After a long and unsuccessful search it was noticed that the dog had a sneaking, hang-dog expression about the mouth, and did not seem inclined to wag his tail as numerously as usual. Somebody suggested that he might have swallowed the bill, and then it was remembered that he had been suspected of that sort of thing before. An emetic was administered, but the dog only howled and gagged, and stubbornly refused to show the "true inwardness" of his disposition. Another emetic was sent for, and finally the missing bill was vomited up with a small tempest of howls and gurgles, and penitential tears. Mrs. X. took the money to a bank and represented that it was soiled and ragged because it had been carried in somebody's pocket for a long time, and the bank was persuaded to take the bill at a liberal discount upon its original value.

Two or three days ago a bill was placed on the floor where the dog could see it, and then everybody left the room, expecting when they returned to see the dog and bill consolidated. But lo! the bill remained untouched, and the repentant canine lay curled up in a corner of the room looking as ashamed and miserable as any human being would have looked under the circumstances.—*Detroit News.*

The Biggest Locomotive in the World.

Uncle Dick weighs sixty-five tons, and is sixty feet long from his headlight to the rear end of his tender. He is the biggest locomotive in the world, and has just been turned out of the Baldwin Locomotive Works for duty on the precipitous inclines of the Atchison, Topeka and Santa Fe railroad. A boiler twenty-one feet long supplies steam for cylinders twenty inches by twenty-six, and gives motion to eight forty-two-inch drivers, while a large tank, surmounting the entire structure, not only carries a water supply, but helps to give Uncle Dick a tighter grip on the rail. His driver will have control over three independent systems of air-brakes, and can bring to bear at once upon his wheels a restraining force of seventy-five tons, which is none too large, inasmuch as a "shoe pressure" of 50,000 pounds is required to keep him when standing still and alone on the steep road over the Ruton pass from surrendering to gravitation and rushing down the hill by his own weight. How heavy these grades are can be understood when it is noted that one end of Uncle Dick will often stand more than three feet higher than the other, so that in traveling his own length he will do the work of lifting about 250 tons a perpendicular foot. And yet this monster, rejoicing in his strength, will rush up the flank of the Rocky mountains with ten loaded cars behind him.—*New York Tribune.*

Newspaper Wit.

There is much genuine wit floating in the newspapers, and much that is bogus to the last degree. When the Camden *Post* says "A man's character is like a fence—you cannot strengthen it by whitewash," it gives a genuine epigram. Quite as felicitous, too, is "Otis," in the Cincinnati *Breakfast Table:* "A tack points heavenward when it means the most mischief; it has many human imitators." A bright turn to a familiar quotation is given by the Biddeford *Miniature* thus: "I am thy father's spirit, as the pint flask said to the inquisitive urchin who had been investigating the cupboard." Mr. Talmage having claimed that hell has four gates, the Buffalo *Express* hopes they open outward, so as to give egress in case of fire. The Hackensack *Republican* said last fall, "The leaves of trees, like summer boarders with bills unpaid, take their departure, leaving their trunks behind them."

A TUTOR of a college, lecturing a young man on the irregularities of his conduct, added with great pathos: "The report of your vices will bring your father's gray hairs with sorrow to the grave." "I beg your pardon, sir," replied the pupil, "my father wears a wig."

A WONDERFUL RUNNER.

From Paris to Moscow in Thirteen Days and Eighteen Hours.

[From the New York Times.]

The present interest in pedestrianism, manifested by the gathering of throngs of people to witness walking and running matches, was equally great in the early part of the century, but was then excited by the wonderful and almost incredible performances of one man. This was Mensen Ernst, born in Bergen, Norway, on Oct. 19, 1799. He was the son of a sea Captain, and when still a boy entered the English navy. Although his powers as a runner were displayed even in his boyhood, he had made three voyages to the East Indies, and had completed a three-years' cruise [illegible] before he made a public exhibition of them, which he first did by running a race in London when 19 years old. Though up to this time he had had no training, and had even lacked a pedestrian's ordinary opportunities for muscular development, he was famed throughout the navy as a runner, and wonderful stories as to his extraordinary powers circulated among all branches of the service. At last a heavy sum of money was wagered by the officers of his ship that Mensen would run from London to Portsmouth, a distance of seventy-three miles, in less than ten hours. The feat was attempted and accomplished, the distance being covered by the sailor in exactly nine hours. Shortly afterward he ran from London to Liverpool, a distance of 150 miles, in thirty-two hours. Although thus successful as a runner, Mensen did not quit the sea until after having distinguished himself by bravery in the battle of Navarino, fought Oct. 20, 1827. Soon after that date he became a professional runner, and, after winning a number of lesser matches, was induced to undertake the great feat of running from Paris to Moscow. He started from the Place Vendome at 4 o'clock in the afternoon of June 11, 1831, and entered the Kremlin at 10 o'clock a. m. of June 25, having accomplished the distance of 1,760 miles in thirteen days and eighteen hours.

This feat created a decided sensation throughout Europe, and the employment of Mensen as a courier extraordinary by Kings and Princes became a popular amusement in European courts. He ran from country to country and from court to court, bearing messages of congratulation, condolence, or dispatches of greater importance, and, whenever matched against the regular mounted couriers easily succeeded in beating them. He always carried with him a map, a compass, and as many biscuits and ounces of raspberry sirup as there were to be days occupied on the journey. In winter he took with him a pair of long, slender Norwegian snowshoes, and, in traveling, he always chose the most direct line, turning out neither for mountains nor rivers, but climbing the one and swimming the other. He never walked, but invariably ran, keeping up a long, swinging lope for hours at a time, without rest. His only refresh[illegible] raspberry sirup per day, and two short rests of ten or fifteen minutes each in twenty-four hours. These rests he took while standing or leaning against a tree or other object of support. At such time he covered his face with a handkerchief and slept, and, after such a nap, he would pursue his way apparently as refreshed as though he had slept for hours. He was very uneasy when compelled to be quiet, and could not sit still for half an hour without feeling a sense of suffocation. He was a very amiable man, a favorite with all who knew him, and was made much of by his royal patrons.

In 1833 he started from Munich at 1 p. m., June 6, with dispatches from the King of Bavaria to his son Otto, King of Greece. These dispatches were delivered at Nauplia at 9 a. m. on July 1; or seven days sooner than if they had been sent by the regular post. In 1836, while in the employ of the British East India Company, Mensen was charged with the conveying of dispatches from Calcutta to Constantinople, through Central Asia. The distance is 5,615 miles, which the messenger accomplished in fifty-nine days, or in one-third of the time made by the swiftest caravan. On this wonderful journey he made his way across terrible deserts, awful salt swamps, where, for hundreds of miles, he saw no living being, and through countries whose inhabitants were savage robbers, and who lived in a state of continual warfare. The man seemed invested with supernatural powers, and his fame spread far and wide. One of the most interesting problems of the age being the discovery of the sources of the Nile, it was suggested that this wonderful man be employed to follow the course of the river to its sources, and thus determine them. Prince Puckler Muskan, who had himself traveled extensively in Egypt, and had made several unsuccessful attempts to penetrate to the headwaters of the Nile, offered to defray all expenses and reward Mensen handsomely if he would undertake the trip. The indefatigable runner consented to make the attempt, and on May 11, 1842, set out from Muskan, in Silesia, whence he ran to Jerusalem, where he had some business to attend to. Thence he continued his way to Cairo and up the western bank of the Nile, through Lower and Middle and into Upper Egypt. Here, just outside the village of Sayne, he was seen on the morning of January 22, 1843, to stop and rest, leaning against a palm tree, with his face covered by a handkerchief. He rested so long that some persons tried to wake him, but they tried in vain, for he was dead. He was buried at the foot of the tree, and it was years before his friends in Europe knew what fate had befallen him.

Gambling in Washington.

A veteran gambler of Washington has been complaining to a newspaper reporter, who has, of course, published his complaints, of the degeneracy of these days, meaning that public men no longer fight the tiger with the energy of old. Senators, Representatives, foreign diplomats, and all the rest, used to risk their money liberally; but now, if they lose a few dollars at the table, they leave the game in a very despondent frame of mind; and, if they win a small amount, they imagine themselves favorites of fortune. The late Humphrey Marshall played with coolness, but always heavily, and was generally lucky. Tom Marshall played irregularly, and was apt to lose unless very drunk, when he was often successful. All the Russian diplomats gambled habitually, according to this authority, Bodisco setting so much aside for the purpose, and, when it was gone, he withdrew, saying the appropriation had been exhausted. Thad Stevens was a scientific player, and always with blue chips. He sent for the reciting gambler the day he died, wishing to see him before he "had passed in his chips," and saying his disease, which the doctors were trying to explain, was simply 78 years of hard work. The faro-bankers formerly cleared $100,000 a season. Then it paid to give free suppers and pop unreckoned champagne. Now they got very little; patrons of the table ate and drank more than they left behind. Even in the South, at such places as the White Sulphur, during the summer, the ancient spirit had departed. There was no longer any fun; men have grown poor and virtuous. The venerable faro-banker has well-nigh despaired of the republic; but the respectable part of the community will wish that his despondency may increase with the increase of its cause. The Government may prosper, even if faro doesn't.

PROVERBS IN COOKERY.

BY MISS DODDS, OF PHILADELPHIA.

There is a greenness in onions and potatoes that renders them hard to digest. For health's sake, put them in warm water for an hour before cooking.

The only kind of a stove with which you can preserve a uniform heat is a gas stove; with it you can simmer a pot for an hour or boil it at the same rate for twenty minutes.

Good flour is not tested by its color. White flour may not be the best. The test of good flour is by the amount of water it absorbs.

In cooking a fowl, to ascertain when it is done put a skewer into the breast, and if the breast is tender the fowl is done.

A few dried or preserved cherries, with the stones out, are the very best things possible to garnish sweet dishes.

Single cream is cream that has stood on the milk twelve hours. It is best for tea and coffee. Double cream stands on its milk twenty-four hours, and cream for butter frequently stands forty-eight hours. Cream that is to be whipped should not be butter cream, lest in whipping it change to butter.

To beat the whites of eggs quickly put in a pinch of salt. The cooler the eggs the quicker they will froth. Salt cools and also freshens them.

In boiling eggs hard put them in boiling water ten minutes and then put them in cold water. It will prevent the yelks from coloring black.

Cutlets and steaks may be fried as well as broiled, but they must be put in hot butter or lard. The grease is hot enough when it throws off a bluish smoke.

You must never attempt to boil the dressing of a clear soup in the stock, for it will always discolor the soup.

In making any sauce put the butter and flour in together and your sauce will never be lumpy.

Whenever you see your sauce boil from the sides of the pan you may know your flour or corn-starch is done.

Boiled fowl with sauce, over which grate the yelk of eggs, is a magnificent dish for luncheon.

Tepid water is produced by combining two-thirds cold and one-third boiling.

To make macaroni tender, put it in cold water and bring it to a boil. It will then be much more tender than if put into hot water or stewed in milk.

The yelk of egg binds the crust much better than the whites. Apply it to the edges with a brush.

Old [illegible] by plunging them into cold water before cooking them.

Never put a pudding that is to be steamed in anything else than a dry mold.

Never wash raisins that are to be used in sweet dishes. It will make the pudding heavy. To clean them wipe them in a dry towel.

To brown sugar for sauces or for puddings put the sugar in a perfectly-dry saucepan. If the pan is the least bit wet the sugar will burn and you will spoil your saucepan.

The water used in mixing bread must be tepid hot. If it is too hot, the loaf will be full of great holes.

To boil potatoes successfully: When the skin breaks pour off the water and let them finish cooking in their own steam.

In making a crust of any kind do not melt the lard in the flour. Melting will injure the crust.

In boiling dumplings of any kind put them in the water one at a time. If they are put in together they will mix with each other.

A Postal-Card on Its Travels.

Jules Verne's "Around the World in Eighty Days" will likely be outdone in point of time yet. We were shown a postal-card yesterday by Mr. Edward G. Hofmann, at the Monongahela House, which is a curiosity, and, if no accident occurs now, will have made the trip around the world in seventy-six days. The postal was sent by Mr. Emil Neef, editor of the *Tagleblatt*, in Zofingen, Switzerland, to a friend, Mr. Leon Roulet, in Calcutta, India, on the 23d of January, with a request to send it from there to Mr. Hofmann, of Pittsburgh, without losing any time, so that it can be ascertained how long it took it to go around the world. It arrived in Calcutta at 10 o'clock a. m., on the 16th of February, consuming twenty-four days on its trip there. In 15 minutes thereafter it was given to the post again, and by mail steamer brought across the Pacific ocean, and, via San Francisco, it reached Mr. Hofmann yesterday, taking thirty days to come here from Calcutta, and sixty-two days since it left Switzerland. It was immediately re-directed to Mr. Neef. It is expected that it will reach him in, at the furthest, fourteen days, as that is the regular time, and sometimes it takes less time for a letter to get there from Pittsburgh. Thus, if no accident occurs now, it will have gone around the world in seventy-six days, four days less than Jules Verne's "Phineas Fogg" required.—*Pittsburgh Gazette.*

Catching Cold.

"Colds" are among the unsolved medical problems. They used to be thought to be due to the suppression of the excretions of the skin; but this takes place whenever the surface is exposed to cold, and often without harm, and colds are sometimes taken when only a few square inches of surface are exposed. It is a fact, too, that men and animals may be varnished without producing the symptoms of a cold.

Still, the ordinary medical view is that, the passage off of effete matter being checked, the blood is altered in character. The corrupted blood then in its turn affects the heat-regulating apparatus.

A cold is a slight fever. It begins with a chilly sensation, followed by heat. The fever runs its course in a day or two. Like other fevers, however, it may have various complications. Hence rheumatic pains, headache, nasal catarrh, sore throat, catarrh of the intestines, herpes labialis—eruptions around the mouth.

Sweating, whether by medicines administered internally or otherwise, is the main reliance for hastening a cure. But the pores should be kept somewhat open by warm clothing, or the heat of a warm room, for several days, during which there should be no exposure.—*Youth's Companion.*

MR. FINNEY, a London dentist, claims to have found a filled tooth in the jaw of an Egyptian mummy. Dentistry was further advanced 4,000 years ago in Egypt than is supposed.

FARM NOTES.

A GILL of raw (unboiled) linseed oil given to a sow, just before and after farrowing, will prevent the disposition to eat her offspring. This habit is an acquired one, and is caused by keeping the sow with others of her own species, who worry and annoy her. Perfect quiet and isolation are considerations which should not be neglected.

WE ferment manure for the same reason that we decompose bone-dust or mineral phosphates with sulphuric acid and convert them into superphosphate, or for the same reason that we grind our corn and cook the meal. These processes add nothing to the amount of plant-food in the bones or the nutriment in the corn. They only increase its availability. So in fermenting manure. This process renders the plant-food in the manure more immediately available.

THE farmer who is habitually careful with his seed-grain, so as to get only the best, is doing much toward advancing the products of his fields, not only in starting his crop—in which it is so important—but throughout its entire growth, including the berry which is to form the seed for the succeeding year. This is apt to be plumper and sounder, and much the better calculated for the next sowing, and an improvement upon the last. I include with the grains the seed also of the forage plants. Always the best growths should be selected, the greatest difficulty being to obtain clean seed, and keep it clean, on account of the foul condition of the land.

AN enthusiastic breeder of poultry says: "Farmers will feed a bushel of corn to produce six pounds of pork, while the same amount of corn will keep a good laying hen one year, and she will produce at least twelve dozen eggs, averaging 18 cents per dozen, which would equal $2.16; in addition she would rear a brood of chickens worth as much more, making a total of nearly $5 against six pounds of pork at 10 cents, equaling 60 cents; or, in other words, the hen will yield seven times as much as the pig on the same quantity of food." Aside from the profit, which is an important consideration in the keeping of poultry, it affords a healthful recreation and enjoyment to all classes. Those who daily toil in the counting-house, factory or mill, each and all need some light, attractive, health-producing employment.

COWS.—Treat them generously and kindly, but do not keep them fat, unless they are to be turned off for beef. A cow is a machine, a laboratory for converting raw materials into milk. If little be given, little will be received. All animals should have exercise, especially those kept for breeding. Some of them are naturally lazy, but they will be the better for stirring about in the open air. It is cruel to keep animals tied up or shut up for days at a time. They need sight, too. Direct sunshine exerts a powerful influence for good on animals, as well as on plants. Do not overlook a good supply of pure water two or three times a day; or good [illegible] When the ground is frozen and covered with snow it may be well enough, on pleasant days, to scatter the fodder, and allow the stock plenty of room to pick it up; but when it is muddy no one but a sloven will fodder on the ground. Good racks should be made, for the sake of convenience and economy.

REMEDY FOR ROUP.—To cure the roup, when a bird is attacked with the characteristic cough of the malady, or has tenacious mucus about the beak with difficulty of breathing, I place it in a wicker coop in a quiet shed, and put before it a drinking fountain containing about a gill of water with which I have mixed one drop of solution of aconite, third potency (may be had of any homeopathic physician). In every instance during three years this treatment has had an effect almost marvelous, for, upon visiting the patient an hour or two afterward, I have found that the symptoms have vanished. The attack for a day or two is liable to return, yet each time in a lighter form; but continuing the application has in no instance with us failed completely to remove the ailment in about forty-eight hours. In case the disease should have made so much progress before it is observed that the sufferer is unable to drink, it will be necessary to give the dose. This is easily accomplished by pouring into the throat about a teaspoonful of the water every hour. Such an instance occurred here during excessive wet weather, when I was absent from one of the houses two days. Upon going to see that all was kept in condition, I found a fine old fellow under one of the porches almost dead from very acute roup. I separated and dosed him immediately. He soon lost all the roupy symptoms, but continued extremely weak, and appeared to be fast sinking from atrophy. A medical friend suggested trying the homeopathic solution of *arsenicum*. His advice was taken, with the best result.—*Fanciers' Journal.*

Pete's Revenge.

Pete McCartney, formerly a very successful counterfeiter, has for three years been a convict in the Illinois State prison, where he is to remain seven years longer. On being convicted, he gave $10,000 to his pretty wife, to live on until he could rejoin her, and she promised to wait faithfully for his release; but she lately made up her mind to get a divorce from him and marry Dr. Mason, with whom she had fallen in love. McCartney learned of her inconstancy, and determined to have his revenge. He employed accomplices to involve the physician in a charge of passing counterfeit money, and then went on the stand as a witness for the prosecution with a story of former partnership in the printing of bogus treasury notes. The prisoner only got out of the scrape by bringing a score of patients to prove an alibi.

Jersey Mosquitoes.

"Was ye iver in Jersey in the summer sason?" said an Irishman to a friend. "'Pon me soul, I was onct, an' I thought I wud be davoured. It was a hot night, an' I raised me winder to catch a whiff of braze, an', instead, I was attacked by a million of murderin' divils wid prongs in 'em. It was a slap here an' a kick there an' a scratch iverywhere. When complately exhausted wid 'em, I giv' up me body for to feast the cannibals, whin, bad luck to thim! the murderin' imps commenced singing in me ear, 'We won't go home till mornin'.' An' well they kept their word, bad luck to thim!' sez I to mesel', as I carried me skeleton out o' the house in the mornin'."

Fatal Tight Boots.

The agony of tight boots is traditional, but so serious a result of the torture is not often reported as the Elmira *Gazette* announces: A lad named Thomas Mear wore a pair of boots that gave as much pain to his feet as they did pleasure to his eye. They not only pinched his feet, but bruised the skin in places, producing an eruption which rapidly took a malignant form, and, going to his heart, produced death.

JEREMIAH GURNEY PHOTO
DAVID S. DIGERNESS COLLECTION

THE GEORGETOWN COURIER for many years had a larger format than "The Denver Post" of today. The page size was 22x28 inches and consisted of nine columns. Jesse Randall's "Courier" front page of June 12, 1889, is reproduced here much-reduced in size.

JESSE SUMMERS RANDALL — on the right — posed with his sister, Iantha Emerine, and his brother, Charles Munro, in 1856. Charles later published a newspaper in Silver Plume, while Iantha cared for their father, Abram, during his later years at Georgetown.

DAVID S. DIGERNESS COLLECTION

MINERS WERE BUSY sacking ore—which could weigh 85 pounds per sack, and be worth approximately $800. Burros then would be loaded and the ore transported to the nearest railway for shipment to the smelter.

Chapter **III**

The Magic of That Precious Metal — Gold!

MINING for the precious metal — gold — is known to be as old as the first gathering of the human race into nations — the earliest approaches toward civilization. The goddess and base of all wealth — gold — has ever attracted all mankind.

It was not long after Jesse S. Randall came to Colorado that he became well-versed in the early history of Colorado, and, to a much greater degree, in that of The Mineral Belt area and vicinity, including its main trading point, Denver City. In his records is the tale of man's attraction to Colorado — that of riches.

The first white man to set foot in Colorado, Francisco Vasquez de Coronado, was disgraced in Spain for failing to find the riches of gold that the Indians claimed existed there. Eighty years before the Pilgrim Fathers landed at Plymouth Rock, Coronado, the last of the great Spanish conquistadores — in gilded armor and waving plumes — led his silver cavalcade through the Golden West, their path a crimson conquest as they sought the legendary seven fabulously rich cities of *Cibolo.* Coronado, crossing the Arkansas River, in the southeastern part of the state, named the country *Colorado,* meaning "color red" in Spanish.

Five flags have flown over Colorado: those of Spain, France (which claimed this entire region as a part of the great Province of Louisiana), Mexico, the Republic of Texas, and the Stars and Stripes of the U.S.A. The first American flag to fly in Colorado was carried by Lieutenant Zebulon Montgomery Pike, who was sent out by President Thomas Jefferson to explore this portion of the Louisiana Purchase, which had been bought from France in 1803. While camped near the present site of Canon City in February of 1807, the desperate party of 24 men, several of whom had frozen feet, were arrested by Spanish soldiers on the charge of being an armed force in Spanish territory, and were taken to Santa Fe. There, Lieutenant Pike met James Pursley, a Kentuckian, who just three years before — while trading with the Indians on the South Platte River — had been taken prisoner and held because of his discovery of float gold in *Bayou Salado,* now known as South Park. James Pursley's steadfast refusal to lead the Spaniards to the spot can be said to have saved Colorado for his country. His was the first recorded gold discovery in The Mineral Belt, as well as in Colorado. However, the word failed to reach any further than Pike's journal and it took another 50 years before gold was to attract the fortune seekers in great numbers to this vast and beautiful wilderness.

Two of the greatest events in making the United States the richest country in the world — as well as a powerful continental nation — were the discovery of gold in California by Captain John A. Sutter in 1849, and, nearly a decade later, the discovery of gold in Colorado by William G. Russell. The California gold strike — known as the California gold rush of 1849 — started a steady stream of travel across the continent, and resulted in the opening of thousands of mines that contributed billions of dollars to the wealth of the nation. In the summer of 1852, a Cherokee cattle trader by the name of Parks, with a party of his tribe on their way to California, discovered gold on the old Cherokee Trail in Colorado, where Ralston Creek empties into what was then known as the Vasquez Fork of the South Platte River — later renamed Clear Creek. A small amount was carried to Georgia, where a portion of the tribe was living. Exhibited in the states through which the party passed, rumor magnified the extent of the "diggings" in the country that was generally known as the Pikes Peak Region. W. G. Russell, who had considerable mining experience, organized a party of 10 men, and left Georgia on February 9, 1858, arriving at the mouth of Cherry Creek on June 24, 1858. Exploring the area, he discovered "color" on Dry Creek, five miles from its confluence with the South Platte, and seven miles south of what was to become Denver. Russell panned the first paying dirt in the territory, yielding $3.00 in pure gold. Returning to Georgia in the autumn of 1858, he carried with him specimens of the precious metal and the event resulted in Colorado's great Pikes Peak gold rush of 1859.

Meanwhile, a man by the name of George A. Jackson, a '49er, left the mines of California in 1857 and returned to his home in Missouri. In the spring of 1858, he came to the Pikes Peak Region more with the view of hunting and trapping than searching for gold. Arriving in the region on the west side of Cherry Creek, he camped near John Simpson Smith's trading post. John Smith and his Indian "Pocohontas" of the Rocky Mountains were the first to build on the site of Denver. He was a trapper and trader, and his Sioux wife, Wapoola, had deserted her people to share in the perilous life of the white

A. E. MATHEWS SKETCH
DAVID S. DIGERNESS COLLECTION

IN THE YEAR 1865, Golden City was the territorial capital and competed with Denver for a place on the Union Pacific Railroad. W. A. H. Loveland planned to make Golden the terminus of his Colorado Central Railroad, which caused a disturbance in Denver. It was known that Loveland had business connections with the Union Pacific. Contruction commenced in Golden on January 1, 1868.

frontiersman. Smith, who later became associated with W. G. Russell and his party of gold-seekers, built a tepee within a picket enclosure during the fall of 1857, in the area that later became Auraria.

During the summer of 1858, George Jackson got the gold fever and tried his luck on the Cache la Poudre River to the north of the original gold strikes. With an old trapper, he prospected for gold there and founded a trading post called LaPorte. In August, he did some prospecting along the St. Vrain and Vasquez forks of the South Platte. (Vasquez, as previously mentioned, later became known as Clear Creek.) In the winter of 1858-'59, Jackson — along with Tom Golden and Jim Sandars — established a base camp near the mouth of Vasquez Canyon (Clear Creek Canyon) upon the site now occupied by the town of Golden — so designated for his companion of that name. From this point, Jackson traveled southward along the foothills to Lupton's Fork (now known as Bear Creek) and northward to Boulder Creek. Then he accompanied a small hunting party, which traveled into the mountains via Mount Vernon Canyon, reaching a beautiful park high on the mountain. This wonderful park proved to be the winter feeding grounds of a large herd of elk and later became known as Bergen Park. They pursued the elk to the north, which lead them to the brink of a cliff overlooking the deep and beautiful valley of what was then the Vasquez Fork of the South Platte River — near present-day Floyd Hill. At the bottom of the canyon they saw the river — frozen solid. In Jackson's time only wild game trails were in evidence; however, a network of highways radiates in almost every direction today.

Separating from his hunting companions, Jackson started out alone, resolved to explore the canyon. Descending a deep ravine covered with deep snow, he floundered his way down to the level of the stream. From this point, he followed its course upstream to Grass Valley, the site of present-day Idaho Springs. Attracted by a white cloud rising from what appeared to be an Indian campfire in a side gulch, he discovered hot springs of soda water, steaming in the cold winter air from the very bowels of the Earth. This was at the mouth of a small creek — later named Soda Creek — which emptied into the main stream of Clear Creek from the south. Jackson observed a herd of Big Horn sheep congregated on the hillsides near the hot spring. He shot one, which supplied him with fresh meat. Years later, a resort area formed at this hot mineral spring,

DAVID S. DIGERNESS PHOTO

THE TOWN OF GOLDEN appears in this panoramic aerial photograph, looking toward the north. The purple peaks in the far distance are just west of Boulder, Colorado, and to the right are North Table and South Table mountains. In the foreground, the remains of inactive clay pits scar the landscape. Fireclay was taken from these pits for the local brick-making industry.

utilizing these health-giving waters.

Prospecting in this vicinity afforded little satisfaction, so Jackson went up Clear Creek barely a quarter of a mile to another stream, and camped just above its confluence with Clear Creek. He built a roaring fire, which thawed the ground and enabled him to dig with his hunting knife. Hunter-turned-prospector again, he used a large tin drinking cup — a somewhat crude substitute for a gold pan — to prospect for any signs of color, the evidence of which every cup imbued him to continue his quest for the precious metal —— Gold! Gold on Clear Creek!

First Gold Discoveries

After digging and washing for some time, Jackson found himself with about $9.00 in gold dust, as well as a fine gold nugget. Convinced that he had made an important discovery, he marked the spot so he could easily identify it, and drew a map in his diary of its location, and he then returned to Golden, after being gone from his base camp 19 days. In his diary, he recorded the date of his discovery as January 7, 1859. During the remainder of the winter, he bought tools and supplies for mining, and had staples shipped in from the East.

While waiting for the heavy winter snow to melt from the gulches, Jackson organized a group of 22 Chicago men to accompany him. Then, on April 17, they moved out with teams and wagons, cutting their way across the mountains through heavy timber to reach the diggings Jackson had located early that year. In many places, it was impossible to move ahead with the wagons, so they unloaded and disassembled them, carried them over the obstruction to a point where they could continue, reassembled and loaded again, and the expedition proceeded until it became necessary to repeat the laborious process. At one point, they unloaded

DAVID S. DIGERNESS COLLECTION

GEORGE A. JACKSON was the discoverer of the first gold in Colorado of large-enough quantity and high-enough value to be commercially worked by the placer method. The famous discovery was made at the confluence of Chicago Creek and Clear Creek, within the town limits of Idaho Springs.

the wagons and lowered them over a cliff, and then loaded them again at the bottom. After experiencing the long, fearful journey to reach the bottom of Clear Creek Canyon, the going became relatively easy. They reached the creek beside which Jackson's diggings were located — naming it Chicago Creek, in honor of his associates from Chicago — on the first day of May, 1859. They turned the wagons into sluice boxes, and mining in the area got underway. In the first seven days, they washed $1,900 from Jackson's sand bar.

Jackson recovered a large amount of gold from the bar and took it to Auraria to pay for added supplies, as well as for provisions belonging to discouraged immigrants and prospectors who were about to return to the States. But, by paying for the goods in gold dust, it soon became apparent that a large discovery, or find, had been made. Jackson was then watched constantly. He finally consented to reveal his secret.

Prior to his trip back to the discovery site of April 17, he had met John H. Gregory, to whom he related information about his discovery on Vasquez Fork, or Clear Creek. Gregory agreed to join him there, and it was while attempting to reach the point designated — as Jackson later related the incidendent — that Gregory, mistaking the direction, followed the north fork instead of the main fork, and ended up with his great bonanza-lode discovery, which eventually made him a fortune.

The Gregory lode originally was called surface gold because it was exposed to the surface. However, it later turned out to be the first great lode discovery in the Rocky Mountains. It was situated in what became Gilpin County, over the hill from the Jackson placer diggings in what became Clear Creek County — which was instrumental in Gregory's discovery. With a crudely constructed sluice, he took out $972.00 worth of gold from the morning of May 16, 1859, through the evening of the 23rd. After working a week, he sold his discovery for $22,000, from which production of more than $20-million was later realized. It is true, as the story goes, "Gold is where you find it," and this saying has many times been repeated, by the prospector and every seeker of gold, whether in the native form, or combined with other metals in ores of great variety.

It is interesting to ponder about the possibility that the finger of fate must have guided Jackson and Gregory in their search for gold, as these discoveries rank among the greatest events in Colorado history — and both Colorado and the United States as a whole have profited. That in two sections of this rugged mountain country — which makes up part of The Mineral Belt — these men should have come almost directly to these particular localities and virtually struck the heart of two of the richest and most extensive mineral deposits ever found, is indeed extremely remarkable. Jackson made his strike while out on what was mainly a hunting trip and without even a gold pan or shovel, and Gregory was on his way overland to California when he chanced on his find — both men acting alone.

Gregory arrived by wagon train at Fort Laramie, Wyoming, and journeyed south to prospect the east slope of the Rocky Mountains. Fate caused him to meet Jackson, whereupon he learned of Jackson's diggings. Shortly thereafter, Gregory headed up in "the hills," where he took the wrong fork of the stream, and almost immediately made his stupendous discovery.

Pikes Peak Gold Rush

The stampede was on, as news of Jackson's discovery, as well as word of the new discovery at Gregory's diggings, spread like wildfire among the discouraged "Pikes Peak or Bust" prospectors at Auraria and back East. There were about 300 miners in the vicinity of Jackson's discovery during the summer of 1859. Placer claims were staked on several sandbars up and down the creek. Sluice boxes, longtoms, rockers and a system of ditches to divert water to the primitive mining devices were located all the way from Fall River down Clear Creek to the foot of Floyd Hill, five miles below Chicago Creek. The main camp was at Jackson's Bar and Payne's Bar, renamed Sacramento Mining Camp because the miners saw a similarity between this new location and the California camp of the same name. Camps were established all along the valley at Spanish Bar, Fall River and Mill Creek and added to the number of other mining camps already booming, such as Sacramento, Grass Valley and Masonville. The several distinct camps in the vicinity of Chicago and Soda creeks — in Grass Valley — were later absorbed by Idaho Springs. Each gold mining camp had to establish its own code of law and local governing bodies. The miners around Sacramento Bar organized the first mining district in what became Clear Creek County in the spring of 1859. This district included the junction of Chicago Creek and Clear Creek, the spot where George A. Jackson made his discovery of gold, that cold winter day of January 7, 1859 — the first gold in Colorado of quantity and value to be commercially worked by placer operations. The gold during the time and period of Jackson was practically pure, metallic yellow in color, and in shape was flour, leaf flake and nugget. At times it was crystallized. All of it was obtained from the placer workings of the stream beds, benches and bars — in gravels and sands from the surface to bedrock.

The "Pikes Peak or Bust" stampede contained many persons who were happy-go-lucky. Some of them were glad to be running away from the debts which had overwhelmed them after the Panic of 1857. Some were always trying to be the first into an area and always had to move every time they heard of a new frontier. All had ideas of getting rich overnight. It no doubt was a strange procession of fortune-seekers of every caliber who pressed in reckless haste on the trail of W. G. Russell. Few artists ever pictured this wild procession, storming so fiercely into these newly discovered mining camps. History has kept scant records of the thousands of men who gained wealth in the Colorado mining industry following Jackson's yellow-metal discovery in what later became Clear Creek County.

During the spring and summer of 1860, prospecting and mining were pushed at an extremely vigorous pace, and many new discoveries of gold-bearing placers and lodes were made from the Saint Vrain to the upper Arkansas River, along the Front Range, encompassing The Mineral Belt. Newcomers from the East flowed into the area, many new sections were explored and many rich discoveries were made. Groups of prospectors crossed the Park Range — better known today as the Mosquito Range — where they began mining in earnest, from the headwaters of the Arkansas River tributaries to the eastern slope and the Blue River tributaries on the western slope. California Gulch — near the site of present-day Leadville — was washed for gold, and began to yield its millions. In present-day Summit County, Georgia, Humbug, Galena, Illinois and French gulches, the continuing gold rush produced wealth which astonished the country. Other exploring parties crossed the Sawatch Range, which forms part of the main crest of the Rocky Mountains — the great Continental Divide — and entered the Gunnison region.

In the area of present-day Boulder County, several districts became very productive, noticeably that of Gold Hill. Mining was exceedingly lively at the diggings in Gregory and Russell gulches in what later became Gilpin County, especially after the introduction of a reliable water supply by means of a consolidated ditch. By this time, thousands of men were engaged in mining in this area, working both gulches and lodes. Quartz mills, 60 in number, mostly of small dimensions, were brought in and set to work. Thirty arastras were also employed, and it was reported that they served well as long as the soft surface dirt was obtainable. Mining affairs also were lively in Clear Creek and its tributaries. In South Park, on the headwaters of the South Platte River, the gold yield was large and general prosperity prevailed.

The gold fever struck Santa Fe Trail travelers at this time and caused them to branch northward, following the Fountain River to Colorado City — the location of the state's first territorial capital and today, part of Colorado Springs' western edge — thence over famous Ute Pass to the mountain diggings in South Park and other areas in the southern portion of The Mineral Belt.

In the spring of 1860, the owners of the Pikes Peak stage line established what was known as the Pony Express. This served as a daily fast-mail service between the Missouri River and the cities of the Rocky Mountain West and the Pacific coast, and was of great value to businesses before the construction of the overland telegraph lines. The scheme was a marvel of American enterprise and served The Mineral Belt, providing faster mail service to the mining camps. Previous to that time, over three weeks were required to convey mails by steamer from New York to San Francisco by way of Cape Horn. The Pony Express covered the distance between the railway terminus on the Missouri River and the Pacific in eight to nine days. Shorter time was required getting it to Denver City and on to some of the mining camps. The Pony Express carried the news of President Lincoln's election from St. Joseph to Denver — nearly 700 miles — in 69 hours, the last 10 miles having been traveled in 31 minutes. A letter by Pony Express cost $5.00. Brave men and first-class stock were required, for Indians and highwaymen often were encountered, and relay stations were sometimes burned and stock run off. Almost the entire distance traversed — nearly 2,000 miles — was one vast wilderness. No delays were permitted, and the mail bags were kept on the move during the whole time of these long, lonely trips. Horses were changed at every station and riders at intervals of from 50 to 70 miles. The rapid time made by the Pony Express caused the government to send the mails overland, and subsequently resulted in the overland stage, which was followed later by the railway. The Pacific Overland Telegraph Company, coping with the exigencies of the Civil War, strung a line from Omaha to San Francisco in 1861. Denver's messages, for a time, still had to be relayed by Pony Express from Julesburg, as the telegraph wires nowhere else touched the new territory. The first wire into Denver, on October 10, 1863, extended to Fifteenth and Market streets. Amos Steck, the mayor, received a congratulatory message from the mayor of Omaha. From Denver, the message rate for 10 words was $7.00 to St. Louis and $9.00 to New York or Boston.

During the summers of 1860-'61, there were busy, bustling mining camps on the headwaters of the South Platte, and portions of South Park were alive with prospectors and miners. These new towns and mining camps flourished while

GEORGETOWN COURIER ENGRAVING

"PROSPECTOR PANNING FOR GOLD" was the title of this "Georgetown Courier" sketch. The prospector was supposed to be at Paynes Bar on Clear Creek in 1859—the same place where H. A. W. Tabor made his first mining claim.

DAVID S. DIGERNESS PHOTO

IN THE GARDEN OF THE GODS, this natural gateway frames a famous view of Pikes Peak (to the west). These red and white sandstone formations are located northeast of Manitou Springs, not far from Colorado Springs. Early settlers were awe-inspired at this sight of Pikes Peak rising above the prairie to an elevation of 14,110 feet.

the placers and gulches were producing great wealth. As the mining gave out on these, the men who had located there abandoned them and moved on to newer diggings. In 1862, rich gold lodes boomed in the Buckskin district, and the town of Laurette had a newspaper, a theatre, nine quartz mills, numerous saloons, parlor houses, an occasional variety-show theatre, and all the gambling houses that were necessary to make the fortune-hunter content and happy. In four years, the glory of Tarryall, Montgomery, Laurette, Buckskin Joe and other camps came to an end, and the settlements became some of The Mineral Belt's first ghost towns and few old-timers remained to tell the story of the dead cities of South Park. It was reported that many an oyster can, full of gold dust and nuggets from the upper South Platte, Arkansas and Blue river drainages, left the southern portion of The Mineral Belt in a single day. Diggings near Montgomery yielded an average of a pound of gold per day to the man.

Denver grew rapidly in 1860, and, during the next several years, it was the point of arrival and departure for almost all who came to or left the Pikes Peak country. Many more-permanent buildings started to show up; some were built of brick, including large business houses which were established at the new metropolis. Among these were three banks, one of which was recorded to have charged from 10 to 25 percent interest on loans per month. In 1862, not long after Clark, Gruber & Co. added a coining and assay department to their banking house, the Federal government purchased their establishment, and it was then used as a United States assay and refining office, becoming the first Denver U. S. Mint.

In the thriving mining camps of The Mineral Belt, frontier mining life was represented in all its phases. There were many men of many lands in the crude habitations that lined the hillsides and gulches. The log cabins, tents and occasional frame buildings that made up the mining camps had the appearance of temporary military encampments. Saloons were numerous, theatrical troupes and girls of parlor house madams toured the region, while religious services were often held in the open air, and bands blaring away enticed the miners into shows or gambling houses at almost all hours of the day or night. Paper money was rarely seen; gold was the universal monitary exchange — usually carried in buckskin pouches. When purchases were made with gold dust it was weighed on scales commonplace in all business houses. Gold, as it came from the gulches, was usually valued at about $18 per ounce.

Placers and gulches were termed poor man's diggings, because little or no money was required to test their value or put them into producing condition. Every man knew what his claim was yielding when night came. The expressions of satisfaction or disappointment in those early Colorado mining times, when sluice boxes were cleaned, would challenge the greed of the miser and the disgust of the spendthrift. Many streams were worked to no advantage. Some ground, which could have been operated by long flumes or hydraulics, was divided among too many owners to secure such operations, and the miners were in too much of a hurry to get rich and leave the country to think of joining together for sure success. Consequently, abandonment followed before long, with no further work being done until consolidations permitted mining on a larger scale. However, in the four summers between 1860 and 1863, the amount of gold obtained was very large. Records indicate the yield from the creeks, placers or gulches alone were as follows: Boulder County, $400,000; Gilpin, $2,500,000; Clear Creek, $1,700,000; Park, $1,500,000; Summit, $5,000,000; and Lake, $4,000,000. This is exclusive of lode mines, which in some locations were worked at all seasons of the year. The lode mines in Gilpin County were extremely productive, and, to a lesser degree, this was also true in most other leading districts of The Mineral Belt.

Congress organized the Territory of Colorado on February 28, 1861 — carving it out of four existing territories, Kansas, Nebraska, Utah and New Mexico — comprising the same area as the present-day state. The first territorial governor, William Gilpin, appointed by Abraham Lincoln, arrived in Denver on May 29, 1861. A census taken at the time showed a population of 25,329, of whom 4,484 were female. H. P. Bennett was elected as a delegate to Congress in September of that year, the first Coloradan admitted to a seat in that body. The counties elected full sets of officials and the first territorial legislature was chosen.

In May of 1862, John Evans succeeded William Gilpin as territorial governor, and S. H. Elbert became territorial secretary in place of L. L. Weld. The following September, H. P. Bennett, a conservative, was re-elected to Congress, defeating William Gilpin, a Republican, and J. M. Francisco, a Democrat, who also ran for that office.

The gulches of Summit County were generally very productive, as was California Gulch in Lake County, both in the southern part of The Mineral Belt — having turned out close to $1-million each summer. Some of the Park County placers had started to fail, but gold lodes near Buckskin Joe, Laurette and Montgomery were paying good profits. The same was true of many localities in Boulder County, in the northern part of The Mineral Belt.

Very few of the men who came to Colorado knew anything of mining or milling. They had pretty smooth sailing, however, when engaged in such simple work as placer or creek mining, or while the soft surface dirt or decomposed vein matter held out. When a depth of from 60 to 100 feet was attained on the lodes, great difficulty was experienced by most men in saving

J. E. RICHARDSON PHOTO
DAVID S. DIGERNESS COLLECTION

OLD-TIME PROSPECTORS often washed "color" from the creek beds using nothing more than a pan. For thousands of years, the gold had been eroded by the action of water and weather on mineral veins high on the mountainsides and washed into the streambeds below.

enough gold to permit any profits because of the hard, complex rock in which the gold was found. The soft quartz had been succeeded by ore from which the gold could not be so easily extracted, especially by the amateur mill men then engaged in the business. In many veins, the rich top material had been succeeded by poor rock, and in some places the veins pinched up to nothing. Many believed the quartz had disappeared for good, and sold their claims for whatever they could get. So, in some localities, there was a suspension of work or production for the want of ore, and in others on account of their inability to extract gold. However, there were other claims that had not been worked deep enough to get below the decomposed mineral, and, at all times, there were large numbers of paying mines.

Rock mining — or better yet, lode mining along deep, rich veins in fissures, cracks and crevice formations filled with mineral-bearing quartz or ore matter — required smelters to separate the gold from its companion rock. It was Professor N. P. Hill who came up with the first process to smelter the complex ores and he originally put the process to work in 1868 in Gregory Gulch (Black Hawk), Gilpin County.

DAVID S. DIGERNESS COLLECTION

POWERFUL HYDRAULIC JETS, such as this one in Alma, Colorado, were used to wash gold-bearing gravel down into sluice boxes containing quicksilver [mercury]. Gold dust and nuggets would sink to the bottom as the material was swept along the sluices by the running water.

The Mineral Belt of Colorado secured great attention as a treasure vault of mineral resources, as the following in-depth story illustrates: The belt's metalliferous veins greatly outnumbered those of any other state and many of them were of surpassing richness. Mining outside the placers was termed "true-fissure-vein mining," which consisted of veins of great uniformity and of varying size. In addition to continuous veins, the region boasted huge ore bodies and enormous deposits, which compared with those of any country in size and value. Under the magic influence of these gold discoveries, a spirit of enterprise was engendered, which accomplished results as unexpected as they were grand and wonderful. The fascination and romance attending the search for precious metals transformed The Mineral Belt wilderness into almost an empire, where the presence of civilized man was unknown only a few years before. Gold mining, of course, comes under two heads, "placer" and "lode," while silver mining is included only in the latter of the two. Lode mining is much more important in Colorado's Mineral Belt, since the quantity of the precious metals contained in lodes or quartz was many times that of the gold of alluvial deposits. When the gold districts were discovered, the first work was usually done in the placers and gulches, where the work was more simple and the gold readily obtained. Less money was required than in lode mining, where either quartz mills or a market for ores was a must. Silver lodes usually required still further time for development and for procurement of the necessary reduction works.

Elaboration will be made on almost all the mining districts of The Mineral Belt in these volumes. This will include information about the immense number of metalliferous lodes, or veins, discovered. A large majority of these were silver-bearing, accompanied with lead or copper. The main value of some of these deposits was in the gold they contained, while others carried all of these metals and iron, as well as other minerals and waste materials. What is sometimes termed a fissure vein, and is considered to be identical with a lode, is a body of gold- or silver-bearing quartz or ore filling a crack or crevice of the foundation rock which underlaid a district. These veins varied in width from several inches to many feet. Some proved to be thousands of feet in length, while others stretched out several miles horizontally. They extended downward into the Earth for unknown distances. The ore bodies contract or close up at intervals, but no well defined veins are believed to have given out entirely. Some of them have a perpendicular direction, while others are nearly flat or inclined at varying degrees. Most of the argentiferous mineral bodies of the Park, or Mosquito, Range appeared in the form of egg-shaped or irregular deposits instead of veins. A short distance away, some of the Leadville carbonates were in the shape of deposits, and others were in the form of veins.

Alluvial deposits — consisting of sand, gravel and rock decay or dirt — were produced by disintegration induced largely by temperature changes such as heating, cooling, freezing and thawing. The prying action of ice and plants, and the mechanical disintegration of the siliceous, granitic and other igneous and metamorphic rocks, also produced alluvium. This material was then transported from the mountains by the agency of water and deposited near the surface of the ground. These deposits were handled by one of the methods known as placer, gulch, creek or bar mining; however, to all of these, the term *placer* is loosely applied. Gold disseminated through the deposits and lodged in considerable quantities on the rocky beds of streams. The prospectors washed "color" therefrom by water, retaining it in pans, flumes and sluice boxes — with quicksilver used to capture it where swift running water was utilized. No roasting, smelting or milling was required, as was the case where the gold was found firmly embedded in ore matter of lodes. In operations of great magnitude, large sums of money were often expended in bringing the requisite water supply to the desired placer diggings. Sometimes, water had to be obtained from a distance in order to work these surface deposits. This often required miles of ditches and flumes to be built. Powerful hydraulic jets were also used for the purpose of driving water into the hillsides with great force, tearing them down more rapidly than could otherwise be done. These placers and gulches called for appliances of varying extent and capacity — from those just referred to, down to the ordinary sluice box, the pick and shovel and the still-more-primitive hand-rocker and pan. Creek mining was carried on along the stream, and on the bedrock far beneath the gravel and boulders immediately underlying the water. In gulch mines, a flume composed of sluice boxes was laid in ravines or gulches, extending up to the bank or head of the excavations. Hoses, hydraulics or falling water from a flume above, were used at this point to wash the banks or hillsides down into the sluice boxes. The sluice boxes were constructed from one to four feet high and wide, were many feet in length, and overlapped one another. On their bottoms were fastened strips of board for riffles, or round blocks sawed from trunks of trees, two or three inches thick. These — with quicksilver placed inside — aided in catching the gold dust and nuggets, which the swift-running waters swept along with dirt, gravel and boulders. The gold, being heavier than the other materials, sooner or later sank to the bottom before it reached the end of the sluices. The intention was always to run the flume on the firm bedrock, which lay under the gravel of the hills and streams. Because the ground worked was nearly level, some of the flumes (the long form of sluice boxes) were run for hundreds and even thousands of feet before bedrock was reached. In other places, the fall was great enough that bedrock was reached in a short distance, and it was easier to retain the requisite incline to the sluice boxes. The water was turned off and the gold taken from the sluice boxes two and three times a day.

In the bar and creek diggings, sluice boxes were used, and the stream was turned from its natural channel. Boulders and heavier stones were removed, and shafts or pits were sunk, until the permanent granite, or bedrock, of the country was reached. In such localities, the gold was usually found extending up and down the course of the stream in a narrow strip of ground. Drifts were excavated along this pay streak, and the gravel, dirt and rock overlying the same was raised to the surface or packed in the drift as an adjoining and parallel one was opened. These drifts (horizontal tunnels) were run from six to seven feet high and wide, and were supported by heavy timbers, which kept the gravelly, rock-strewn stream beds — often from 15 to 50 feet thick — from caving in. If no pay streak was found after sinking a shaft, it was prospected for by running drifts out from the shaft, as already noted. Water which continually leaked into the mine from above, was removed by hand power, using buckets. In later mining of this type, pumps were put to use, sometimes of great capacity.

In the last half of the 1870's and into the 1900's, placer operations of great extent were inaugurated by large dredging operations in the southern part of The Mineral Belt — in Summit, Park and Lake counties. There also was some extensive mining of this character in Clear Creek and Gilpin counties. In the same areas, less-modern methods were also used. These were supplied with hydraulics, great flumes and other facilities for handling vast amounts of dirt and gravel. The rapidity with which the ground was worked permitted large profits.

Lode mining was conducted very differently from placer operations. After the decomposed surface material had been passed, gold was not so easily extracted as from pay-dirt and gravel, and silver ores were usually much more difficult to handle. Hard as most ores were, the enclosing rock was still harder, and the miners' difficulties did not end with blasting, breaking and raising the ore to the surface; for then came the reduction of the ore, which called for costly machinery, and

DAVID S. DIGERNESS COLLECTION

HYDRAULIC MINING in the gold diggings of French Gulch in Summit County are shown in this artist's conception. This illustrates two heads of water washing banks down into sluice boxes.

for skill and experience in operating the same, and in running the milling or smelting process in order to secure the desired results.

For the purpose of breaking and securing ore, the requisite entrance into the domains of Mother Earth was made by digging and blasting out perpendicular shafts and horizontal levels or excavations on the vein. This was done by hand or machine drilling, and blasting powder and high explosives were brought into play. As work progressed, solid timbers were required to brace the sides of the mine. These timbers or log supports extended from wall to wall, and kept the walls from falling in. For 50 to 100 feet below the surface, manpower and the windlass were used sufficiently for hoisting the ore, waste-rock and water. After that, horsepower was required, which was provided by a so-called "whip" (or pulley), or more generally by a "whim" (a large drum or

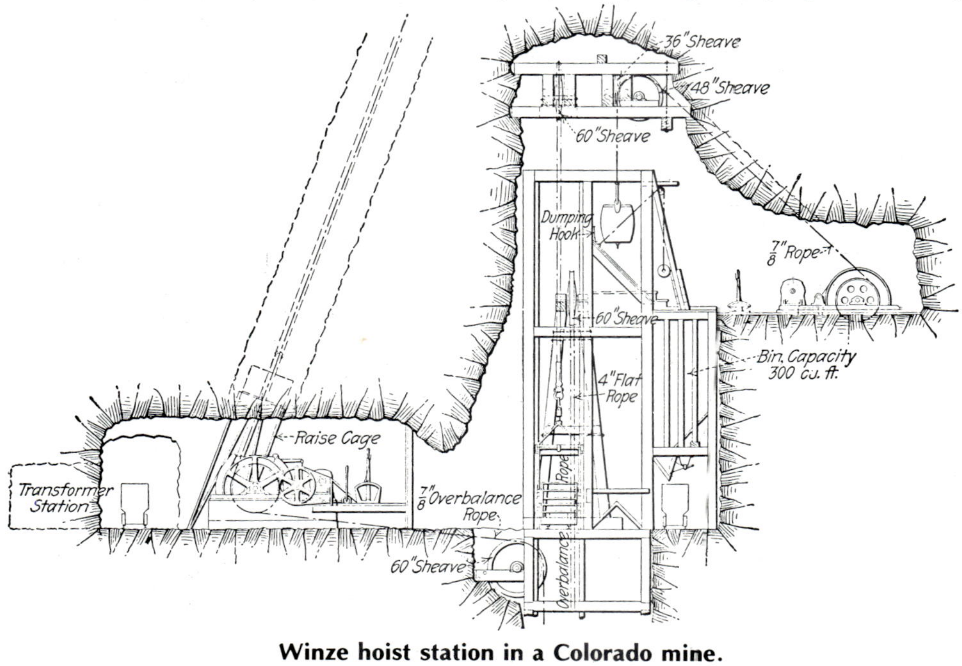

Winze hoist station in a Colorado mine.

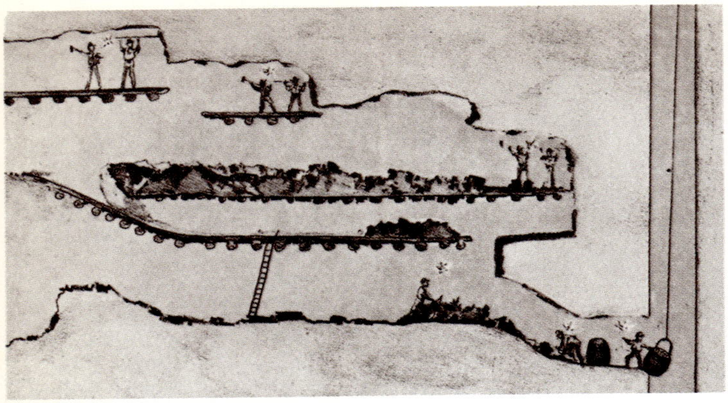

A. E. MATHEWS SKETCH
DAVID S. DIGERNESS COLLECTION

A DIAGRAM of underground mining, this sketch shows shaft-and-mine workings as miners worked on platforms attached to the walls, with poles driven into the rock for support.

timber, around which the rope winds and unwinds as the ore bucket ascends or descends). This usually answered all purposes down to a depth of 200 feet or more; mines of greater depth required the use of steam engines and the accompanying hoisting machinery. Sometimes the whim was dispensed with; steam following manpower and the windlass. When greater depth was gained and the mines were enlarged, there was more ore to handle and the water became more troublesome. More powerful hoisting works were then required, and huge pumps were a necessity. As levels were extended and the vein material was broken down, iron tracks were laid, and the ore was conveyed in cars from various parts of the mine to the shaft, where it was placed in buckets and hoisted to the surface. From there it was sorted and taken to the mill or smelter. There were a number of requisites in lode mining: the development and working of mineral veins; and the treatment of ore after it left the mine — both in stamp mills and reduction works — which will be fully explained later. The operations were much more difficult and intricate than in gulch mining, where the miner could clean up his gold every night instead of waiting for the results of the somewhat tedious process of shipping, milling and smelting. In past ages, gold found its way from lodes on the hillsides down into streams or outlying ground; there it gradually worked its way through the loose gravel onto the hard bedrock, where the richest streals were found. The gold usually was as fine as powder, but often was found in coarser particles and in nuggets, or lumps, of various sizes. It was not free from impurities; and, consequently, it did not come up to the value of artificially refined gold, which was worth nearly $20.67 per ounce in 1878. Placer or gulch gold in Colorado was worth from $17.00 to $18.50 per ounce in 1878.

Although nuggets or particles of gold are often seen in quartz, precious metal is usually disseminated through it, so as to be invisible to the naked eye. Beneath the surface, or deep in the mines, the ore brought to the surface is so hard that it required the aid of ponderous stamp mills — which were driven by water or steam power — to crush it to the requisite fineness for washing or raw amalgamation, or for roasting or smelting. Even then, the stamps did not always pulverize it fine enough for successful raw amalgamation. The mineral resources of Colorado are already known to be enormous, and further discoveries are constantly being made. They constitute one of the main wealths of the state. Gold, silver, lead, zinc, copper, tungsten and molybdenum, along with other minerals and non-metallic minerals, such as coal and clay — not to mention others — have all been extensively mined. Many other materials abound, and some of them will yet cut an important place in the future.

Mining in former years and at present, are two very different undertakings. Most of the men engaged in mining and milling in the very early days of Colorado were novices in the business, and could neither work understandably nor to the best advantage. Labor was usually scarce and high-priced, and much of it of little value. The country was remote from railways and supply centers. Indians at times blocked the routes of travel and impeded the slow-going mule and oxen transportation. It usually cost as much to freight machinery from a railway terminus to the mountain mines as it did to buy it from the manufacturer. Nearly everything consumed was imported from the distant states. In some years, freight charges alone amounted to what would later pay both purchasing and shipping costs. The methods of extracting gold from the ore were at one time so imperfect that the greater part of the treasure contained therein could not be saved. Lode veins were originally divided among too many separate owners, and the claims were too small to work profitably to great depths without steam-powered machinery.

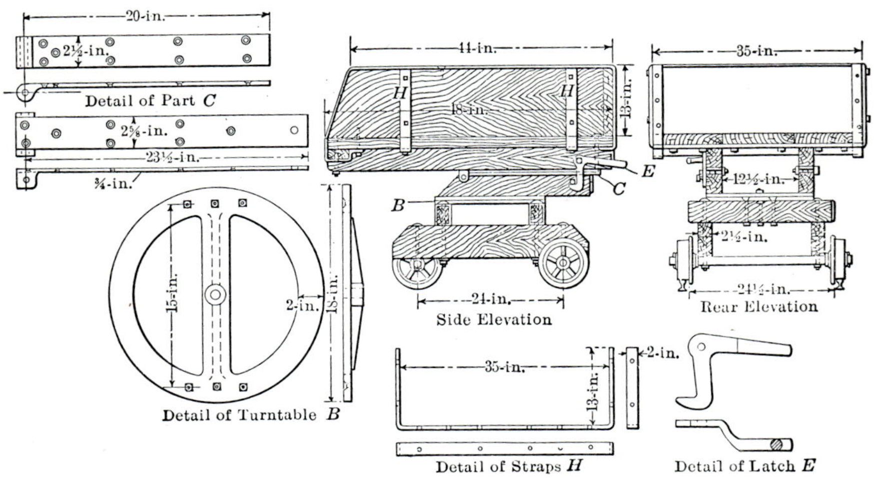

Car for tramming boulders, as made at Joplin.

Consequently, later mining companies required the consolidation of many operations under one management to insure success. A great change was brought about by the railroads as they were built into The Mineral Belt, bringing about cheaper, more-rapid transportation, and by reducing expenditures of labor and supplies. By improved mining and milling, and economical and systematic management, a fair price market for the miners' ore product was in turn achieved. Railroads, such as the Denver, South Park & Pacific, the Denver & Rio Grande and the Colorado Midland penetrated the southern portion of The Mineral Belt. However, the Denver, South Park & Pacific — known as the "Old South Park" — connected the region by way of the shortest route from Denver City, serving the greatest area, as well as penetrating into the Gunnison region.

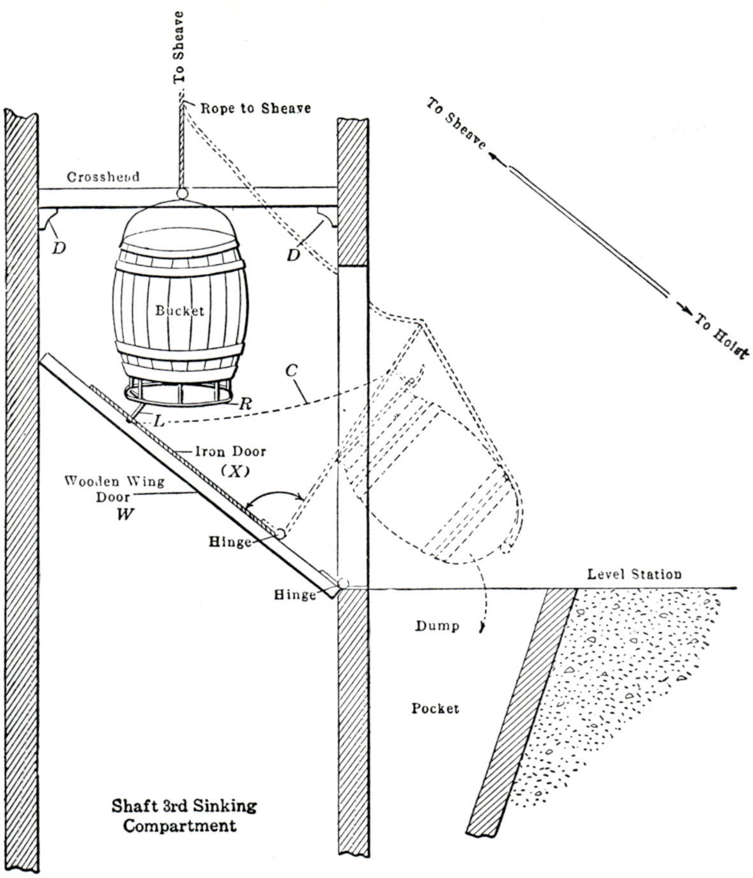

Arrangement used at Kennedy mine for dumping sinking bucket.

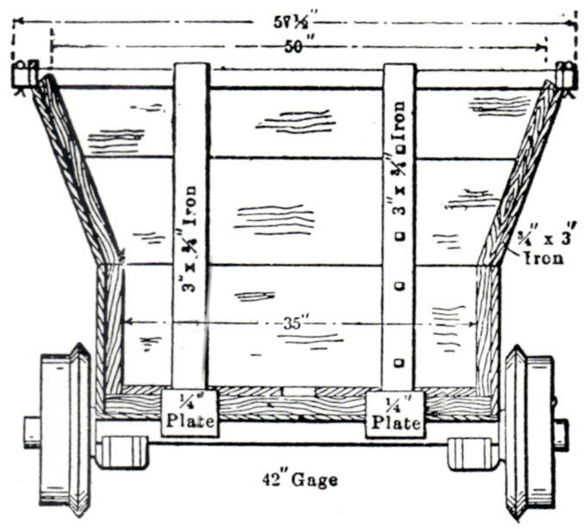

Sulphur, Mining and Railroad Company's ore car.

DAVID S. DIGERNESS COLLECTION

A HORSE-POWERED WINDLASS was one means of raising or lowering ore buckets into mine shafts. A large wooden drum, called a "whim," turned and wound the rope or cable as the horse circled. This line drawing is of the Iron-Silver mine at Leadville.

GURNSEY PHOTO
DAVID S. DIGERNESS COLLECTION

IN THE AFTERNOON, Ute Pass was one-way downgrade—jammed with traffic returning from the mountain mining camps. It might be noted that this mule-hauled wagon train is nearly devoid of freight, suggesting that more was going into the prospecting areas than was coming out. This wagon train of 1878 belonged to the Rocky Mountain Transportation Company. Colorado Springs freight lines had 2,000 animals engaged in moving materials and goods into the mining districts, and yet they lacked the men, wagons, oxen and mules necessary to handle all the business available to them. Later, Ute Pass provided a route for Colorado Midland trains, which gained them access to South Park and Leadville.

Chapter IV

Coming of the Iron Trail

THE MINERAL BELT NEEDED RAILROADS, and three men stand out as the early railroad-construction giants through-and-through. William A. H. Loveland, builder of the Colorado Central Railroad; the Honorable John Evans, builder of the Denver, South Park & Pacific Railroad; and General William Jackson Palmer, builder of the Denver & Rio Grande Railway.

But, first — before we get into that thrilling period that Jesse S. Randall followed closely, reporting on their progress as new segments opened — we will go back to 1862, when President Abraham Lincoln signed the Pacific Railway Act. At that time, Denver was the only town of importance in the Rocky Mountain Region and the people of the city were enthusiastic with visions of railroad communication in the near future. It did not seem possible to them that a railroad to the Pacific could be built anywhere except by way of Denver City. The Union Pacific and Kansas Pacific railroads were authorized, under the Pacific Railway Act of 1862, to build railroads west to the Pacific Ocean.

In 1865, the scheme of W. A. H. Loveland to make Golden a railroad center caused a great disturbance in Denver. Golden City was then the territorial capital, and was boastful and arrogant in regard to its supremacy over Denver City. The competition between the two places became fierce.

In 1865-'66, the construction work on the Union Pacific and Kansas Pacific lines was pushed with great activity. John Evans and other enterprising men began laboring with the Union Pacific directors to bring their road this way. They mapped out a route through Denver, up Clear Creek Canyon, and thence by way of Berthoud Pass, across western Colorado to Utah and beyond. This plan was all brought about by an even-earlier project in 1861, when a survey was made by E. L. Berthoud to determine the feasibility of a stage-and-wagon road through the mountains between Golden, in Colorado Territory, and Salt Lake City, in Utah Territory. The survey also demonstrated that a railroad could be built over the same route. Mr. Loveland, at the time, was engaged in building toll roads, and was the one behind E. L. Berthoud's survey. In the summer of 1868, the news hit Denver — like a thunderbolt — that the mainline of the Union Pacific would miss Denver entirely, passing over 100 miles to the north.

Mr. J. S. Brown is quoted as saying:

Only those who lived at that time could appreciate the danger that threatened the city, people were leaving, going to Cheyenne or some other place along the line of the Union Pacific, which was being pushed west across the continent, linking the East and the West. Business was brisk in Cheyenne and deadly dull in Denver. We were at the tag end. Transportation facilities retarded the development of mines and of agriculture as well.

We sat here during the middle of the 1860's contentedly confident that the Union Pacific would build its main line through Denver. But it did not; stagecoaches dropped down from Cheyenne; we were sidetracked. No one felt certain of remaining here. While the railroad question was in doubt, I was among the many who visited Cheyenne, with a view of locating there. The First National Bank of Denver had a branch in Cheyenne; the Kountze Brothers had a branch there; George Tritch opened a store there. It looked for a time as if Cheyenne would be the one great distributing point of the Rocky Mountain region.

In the evening the men collected around the doorway and in the halls of the old Broadwell house to discuss the situation, and they decided to create a Board of Trade for unity of action. It was organized November 13, 1866.

Meanwhile, W. A. H. Loveland, always enterprising and vigorous, was pushing forward his railroad project to make Golden the chief city of Colorado.

The new Board of Trade called a meeting, and the house was packed with serious, thoughtful people. General Bela M. Hughes addressed the audience in his forceful manner, urging the importance of standing loyally by the organization of businessmen and the imperative necessity for immediate action upon the railroad question in order to maintain the supremacy of Denver.

Colonel Archer of the Kansas Pacific, in a brief speech, stated that the Kansas Pacific would build to Denver if the people would give a bonus of $2,000,000; otherwise it would not. There were many in the audience who had not been told of the proposition that Colonel Archer had been authorized by his company to make, and it fell like a thunderbolt among them, they were simply stunned.

George Francis Train, sent to discuss the railroad question, by General Grenville M. Dodge, realizing that something must be done immediately; followed with a sensible, vigorous and logical address; he outlined a plan for Denver to build a line of her own to connect with the Union Pacific at Cheyenne.

"Colorado is a great gold mine, Denver a great fact; make it

CHARLES WEITFLE PHOTO
DAVID S. DIGERNESS COLLECTION

HEAVY WAGONS, drawn by long teams of oxen, were a common sight on Ute Pass during the rush to Leadville. Terribly slow and expensive, this mode of travel made the efficiency of railway transportation a necessity. Seizing the opportunity, railway companies laid trackage to Leadville from Denver (DSP&P), Colorado Springs (CM) and Pueblo (D&RG). And before the track-laying was over, nearly every town of importance in the Colorado Rockies had rail service. To this day, the principle of "steel wheel on steel rail" forms the basis for a large percentage of Colorado's freight movements—for no other form of transportation uses less energy per ton moved.

a railroad center," said Mr. Train. The town builders were thrilled and filled with courage.

The new Board of Trade immediately acted upon Train's advice. A mass meeting was held in the old Denver Theatre with W. F. Johnson as president; John Walker, of the *Denver Tribune*, secretary; John Evans was the principal speaker. He had such a mighty faith in the future of Denver that one stood appalled at its mightiness. At this meeting he predicted that within fifty years Denver would be the bullion center of America, and it would be made an important railroad center.

Within 20 years, both his prophecies were fulfilled. Talk gave way to work. The Denver Pacific Railway Company was incorporated. The officers were: Bela M. Hughes, president; Luther Kountze, vice-president; David H. Moffat, treasurer; W. F. Johnson, secretary; and F. M. Case, chief engineer. The board of directors was made up of Joseph D. Bates, William M. Clayton, John Evans, Bela M. Hughes, W. F. Johnson, Luther Kountze, David H. Moffat, John Pierce and John W. Smith. All these men were of importance in themselves. "We knew with these men at the front the road would be built," said Mr. Brown in conclusion.

Denver was a little town on the edge of civilization; Cheyenne, Wyoming, as the headquarters of railroad construction, was throwing a long, dark shadow over Denver. Hard times were a reality. Yet in four days, $300,000 in stock had been subscribed. However, this was not sufficient to complete the railroad. Arapahoe County issued $500,000 in bonds. They had difficulty in disposing of them. A committee went to Chicago to sell them, but Chicago bond dealers did not think they were of much value. Arapahoe County was wilderness, and Denver City its main area of civilization. But when John Evans, a bearded man of large build, put his shoulder to the project, it moved. He was determined to go ahead.

The ground-breaking ceremony for the Denver Pacific was an occasion for much rejoicing. Two women, Mrs. Frederick J. Stanton and Miss Nettie Clark, held the plows that turned the sod. A large number assembled to witness it; a brass band filled the air with music, and William Gilpin made one of his famous prophetic speeches.

GURNSEY PHOTO
DAVID S. DIGERNESS COLLECTION

A MULE TRAIN ON UTE PASS belonging to the Rocky Mountain Transportation Company carried covered wagon loads of groceries and merchandise to the Leadville area from Colorado Springs. The wagon road was one-way upgrade in the morning hours.

Lith. by J. Bien, N. Y. — Entered according to Act of Congress, in the year 1866, by A. E. Mathews, in the Clerk's Office of the District Court of the United States for the Southern District of New York. — A. E. Mathews, Del.

BLAKE STREET, DENVER, COLORADO.

A. E. MATHEWS SKETCH
DAVID S. DIGERNESS COLLECTION

BLAKE STREET in Denver was the principal business street for many years. This scene, looking northeast from near the Cherry Creek bridge, shows Blake Street in 1866, when it was one of the busiest trading points in The Mineral Belt. The merchants reportedly sold $12-million worth of goods the previous year. Most purchases were made with gold dust carried in buckskin pouches. The building on the left was a hotel and saloon known as the Elephant Corral, where gambling went on day and night — to the accompaniment of a tinny-sounding orchestra and clinking glasses. Originally called the Denver House, this establishment was referred to as ". . . the Astor House of the gold fields." It was renamed for the large corral that had been between Fourteenth and Fifteenth streets, extending to Wazee Street.

DAVID H. MOFFAT was a prominent railway builder and was treasurer of three rail lines operating out of Denver. Moffat also was vice-president of the Little Pittsburg Consolidated Mining Company at Leadville.

GEORGETOWN COURIER CUT

When the building of the Denver Pacific was an assured thing, the inhabitants of Denver politely informed the Kansas Pacific Railroad officials that they could bring their road into Denver if they so desired, but no bonus would be paid. Then commenced a lively race in railroad building, and the Kansas Pacific reached Denver on August 15, 1870, two months after the Denver Pacific.

It was the custom in those days to paint the name of some individual on the locomotives. The first locomotive, named *David H. Moffat*, arrived with the construction train on June 15. This engine, prior to its purchase by this company, had

Lith. by J. Bien, N. Y.   Entered according to Act of Congress, in the year 1866, by A. E. Mathews, in the Clerk's Office of the District Court of the United States for the Southern District of New York.   A. E. Mathews, Del.

LARAMIE STREET, DENVER.

A. E. MATHEWS SKETCH
DAVID S. DIGERNESS COLLECTION

LARAMER STREET was spelled "Laramie" in this sketch of old Denver City in 1865. A. E. Mathews was looking northeast from Cherry Creek when he created this realistic work of art. Denver had become a major supply center for the mining camps during this period.

GOVERNOR JOHN EVANS, principal builder of the Denver, South Park & Pacific Railroad, also built the standard-gauge Denver Pacific Railway—the first rail line to reach Denver.

GEORGETOWN COURIER ENGRAVING

been known as Number 29 of the Union Pacific Railroad, and had established quite a record for itself. It was the first to enter the town of Cheyenne; the first to cross the Sherman Mountains, west of Cheyenne, as well as the Rocky Mountains; the first to signal its presence in the valley of Salt Lake; and finally — with veteran engineer, Sam Bradford at the throttle — the first to announce to the people of Denver the completion of their first railway.

On June 24, 1870, the first passenger train came into Denver. The Honorable John Evans drove the silver spike, which was a donation from the silver mines of Georgetown. On one side it was engraved, "To the Denver Pacific Railway," on the other side, "John Evans, president."

The event was celebrated with music, processions and speeches. These two railways formed the nucleus around which Colorado's present-day railroad systems were built. Only the group of railroad builders of that time knew of the difficulties, the uncertainties, the haltings, the sanguine hope and utter despair that attended the pioneer railroad

Lith. by J. Bien, N. Y. Entered according to Act of Congress, in the year 1866, by A. E. Mathews, in the Clerk's Office of the District Court of the United States for the Southern District of New York. A. E. Mathews, Del.

F STREET, DENVER.

A. E. MATHEWS SKETCH
DAVID S. DIGERNESS COLLECTION

THIS ARTIST'S RENDITION by A. E. Mathews shows downtown Denver in the 1860's. The viewer is looking up "F" Street — known today as Market Street. Denver's red-light district flourished here for a period when the street was known as Holladay Street. Because of the many parlour houses here, and the reputation of the district, the Holladay family requested that the name be changed back to Market Street.

enterprises. They had many trying situations to meet and great obstacles to overcome; financial and political difficulties clogged their way, and litigations beset them on every hand.

While these men were struggling to build the Denver Pacific, they were also planning the construction of various other roads in Colorado. There were five master leaders among the men who laid the foundation of Denver's great railroad system: John Evans, William A. H. Loveland, William J. Palmer, Jerome B. Chaffee and David H. Moffat. The greatest achievements in early railroad construction are justly credited to Captain E. L. Berthoud. While chief engineer and secretary of the Colorado Central Railroad Company, he designed and engineered the construction, under Mr. Loveland's directions, of standard-gauge lines from Golden to Denver and Golden to Cheyenne, and the famous narrow-gauge lines up the Clear Creek valleys and canyons. He discovered the pass over the Continental Divide west-northwest of Empire, Colorado, that bears his name.

James J. Hagerman, builder of the Colorado Midland Railroad; David H. Moffat, builder of the Denver Northwestern & Pacific Railroad, forerunner of the Denver, Salt Lake & Pacific; Bradford H. Locke, builder of the Gilpin Tramway; Edward John Wilcox, builder of the Argentine Central Railway; and the Union Pacific Railroad, builder of the Greeley, Salt Lake & Pacific Railroad, forerunner of the Colorado & Northwestern and also concurrently, the Denver, Boulder & Western — all stand out as the railroad-building giants in The Mineral Belt's railroad annals.

The Colorado Central

Returning to our story — as Jesse S. Randall closely followed the progress of the building of these railroads and their operations in The Mineral Belt — we will follow his footsteps as he traveled, gathering information and news items. Bit-by-bit, his journals, records and news stories became a comprehensive history of each railroad, starting with the Colorado Central.

The Colorado Central Railroad was an essential co-operator in the development and advancement of the wealth and mining industries of The Mineral Belt. It afforded transportation and traveling facilities for some of the best mining sections. The counties of Gilpin, Clear Creek, Jefferson, Boulder, Arapahoe and Larimer all were traversed to a greater or lesser extent by this road. Until Leadville came to the front, the first two named counties supplied two-thirds of the state's

W. G. CHAMBERLAIN PHOTO
DAVID S. DIGERNESS COLLECTION

DENVER CITY, in Colorado Territory, had this appearance in the late 1860's as the viewer looked toward the southwest, down Blake Street and across Cherry Creek. The famous "Elephant Corral" was on this block, and is shown here with the appropriate sign. Auraria is on the west side of the streambed.

gold and silver export. The other counties were equally conspicuous for their wheat and farming products, livestock, or coal, and other mineral output. A more rapid form of transit, greater accessibility to the outside world, and cheaper freight costs for supplies and goods were the main advantages of this railroad system, resulting — at the time — in additional and/or enlarged mining and business operations.

When it was completed, the Colorado Central became Jesse's main mode of travel to and from Georgetown, whirling him along the enchanting canyon of Clear Creek in luxurious railway cars — at times mixed-in with freight cars of ore and other merchandise headed for their destinations by way of this iron trail of the Rockies.

This narrow-gauge mountain division of the Colorado Central was a marvel of engineering skill and American enterprise, and much of it was constructed under more-than-ordinary difficulties. There was lack of money, opposition from rival schemes, and a belief that the project of building and operating a railroad up such a wild and rugged canyon was not practical. The men of this region, however, had great sagacity and public spirit, which led them to boldly conceive, and eventually carry forward, the undertaking to a satisfactory completion. Ultimately, their efforts resulted in the fruition of most of their dreams. To this day, people still marvel at their creation, in spite of the fact that it is a thing of the past. W. A. H. Loveland and Henry M. Teller — both prominently identified with building the territory and state — were the leading original promoters, and eventually secured the construction of the road from Golden to Denver and from Golden northward to Boulder; this being built as standard-gauge. A narrow-gauge line (three feet between the rails) was pushed west from Golden to Black Hawk. It was an auspicious event when the wild canyons of Clear Creek first echoed to the scream of the steam whistle of those first locomotives, built by Porter Bell — 0-4-0 tender locomotives — thereby providing the mines of Gilpin County with rail communication with the world. This was in 1873, nearly three years after the first track-laying began westward from Jersey Junction on the Denver Pacific and Kansas Pacific, north of Denver City. But, actually, the Colorado Central Railway was started back in 1868.

William A. H. Loveland had retained his alliance with the Union Pacific through the several years of his endeavors to inaugurate construction of the Colorado Central Railroad, and made a nominal beginning of work at Golden City on January 1, 1868. Jefferson County voted $100,000 of bonds in exchange for stock in the road; and, before the middle of the summer of that year, the company had contracted for six miles of grading and masonry work on the first division of the road — extending eastward from Golden. This division did not run a line directly into Denver, but to a junction with the Denver Pacific and Kansas Pacific roads, near the point where the old Omaha & Grant smelter was located, which now is the parking lot of the Denver Coliseum. At that time, it was about

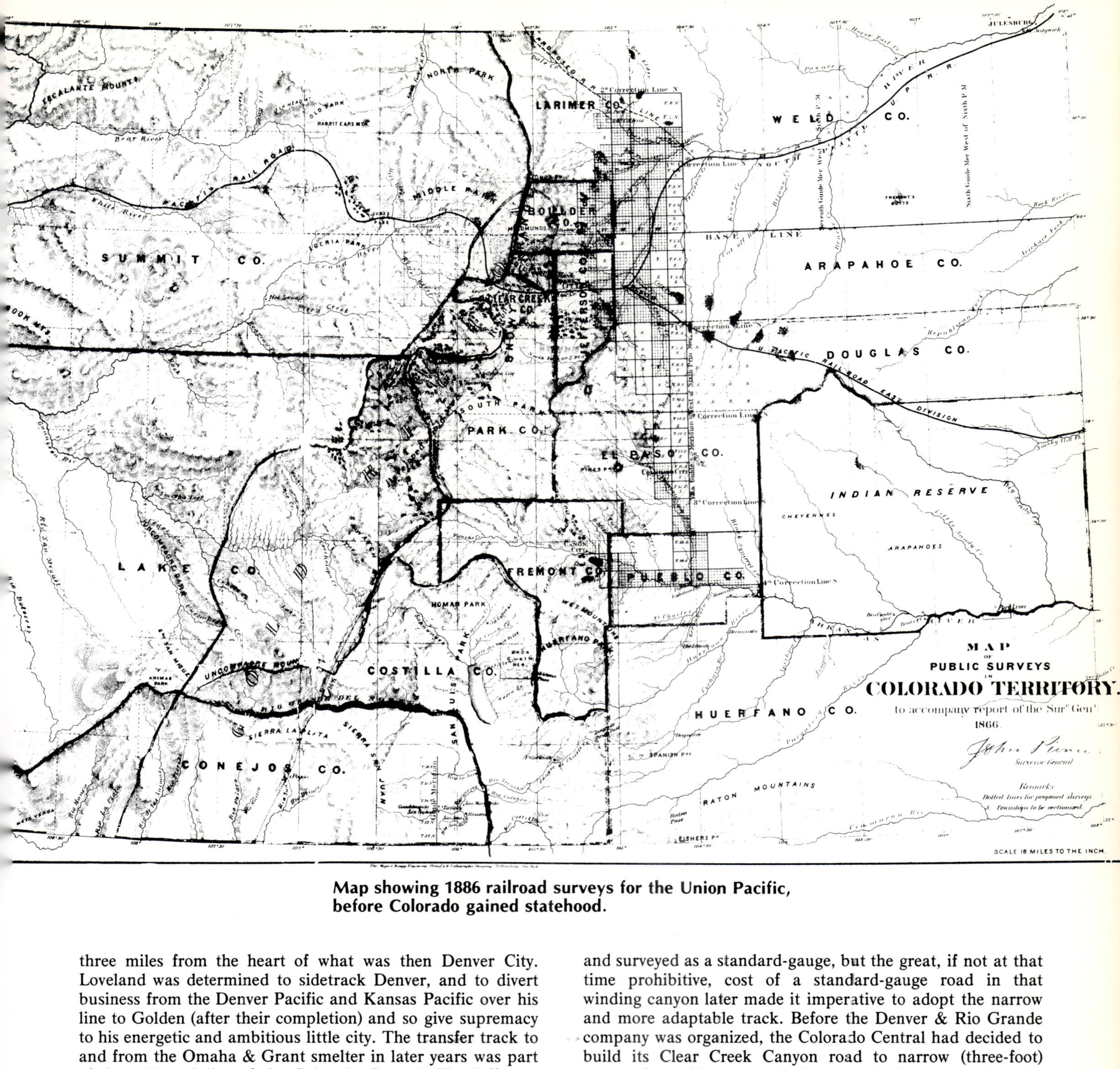

Map showing 1886 railroad surveys for the Union Pacific, before Colorado gained statehood.

three miles from the heart of what was then Denver City. Loveland was determined to sidetrack Denver, and to divert business from the Denver Pacific and Kansas Pacific over his line to Golden (after their completion) and so give supremacy to his energetic and ambitious little city. The transfer track to and from the Omaha & Grant smelter in later years was part of that old mainline of the Colorado Central. The Jefferson County bonds were about the only resources the company had at that time, and by the close of 1868, $87,000 had been expended on the road. Although the enterprise was controlled by the Union Pacific interests, they had not provided any money for its advancement. At the annual election of directors, held January 11, 1869, 6 of the 11 members chosen to constitute the board were Union Pacific men, namely: T. J. Carter, F. G. Dexter, John Duff, A. Lambert, J. B. Taft and J. G. Tappan. The other five were W. A. H. Loveland of Jefferson County, John Turk of Clear Creek County and Henry M. Teller, Truman Whitcomb and E. K. Baxter of Gilpin County. The railroad east from Golden was to be of standard-gauge, while that part of it up Clear Creek Canyon was to be of narrow-gauge. The entire line had been projected and surveyed as a standard-gauge, but the great, if not at that time prohibitive, cost of a standard-gauge road in that winding canyon later made it imperative to adopt the narrow and more adaptable track. Before the Denver & Rio Grande company was organized, the Colorado Central had decided to build its Clear Creek Canyon road to narrow (three-foot) gauge. Some history books have made it appear that the D&RG was the pioneer in preparing for the introduction of the narrow-gauge system in the West; but that is an error.

In 1870, when the Denver Pacific and Kansas Pacific roads were approaching completion, active construction work was resumed on the Colorado Central eastward from Golden and was continued without interruption until the grading of the roadbed was completed to a junction with the DP and KP in September of that year. Trackage and equipment was provided by the Union Pacific, and the first train passed over it between Denver and Golden on September 23, 1870, using the Denver Pacific track from the city to Jersey Junction and the Colorado Central Railroad. The line was later changed (in October of 1874) to a more direct route into Denver City. Upon the opening of the railroad to Golden City, the stage

STATE HISTORICAL SOCIETY OF COLORADO

THE DENVER PACIFIC Railway reached Denver in the summer of 1870, the first line to do so. This photograph shows the standard-gauge locomotive "Walter S. Cheesman" upon its arrival in the "Queen City of the Plains."

lines that had operated between Denver and the Clear Creek mining towns made Golden their eastern terminus.

The Union Pacific people had come to the active support of the Colorado Central in 1870, principally because the Denver Pacific — at one time promised to be their connecting line to Denver and the Colorado mining districts — had become practically a part of the KP, thus compelling the UP to adopt other means for securing the business of the Colorado mining country of The Mineral Belt. An ambitious campaign was planned and opened the way for the Colorado Central Railroad. In 1870, the narrow-gauge line from Golden up the canyon of Clear Creek was started. At the same time, a standard-gauge road was surveyed direct from Golden down the valley of the South Platte to Julesburg, which was intended to be the mainline, leaving Denver off to one side to be reached by a branch road from the junction with the Denver Pacific and Kansas Pacific railroads. This Golden-Julesburg road was located through the Marshal coal fields, Boulder, Longmont and Greeley; and, in the month of July, 1872, Boulder County voted it $200,000 in bonds, and Weld County voted for $150,000. Work on the road began in September of 1872; and, by the middle of the summer of 1873, the track was completed from Golden to Longmont and opened for traffic. The Panic of 1873 stopped further construction on the line and it was never resumed by Loveland. In later years, the Union Pacific (cut-off) road from La Salle to Julesburg occupied part of that Golden-to-Julesburg route; and, still later, the track was taken up from a part of the section that had been completed between Golden and Longmont. The Golden-to-Julesburg line, connecting with the railroad up Clear Creek Canyon, was the main reason for defeat of construction of a railroad direct from Denver across the mountains. This idea of a line direct from Denver was conceived by John Evans. At that time, the rivalry between railroad undertakings represented by John Evans and his

C. DUHEM & BROTHERS PHOTO
STATE HISTORICAL SOCIETY OF COLORADO

THE KANSAS PACIFIC Railway reached Denver on August 15, 1870, two months behind the Denver Pacific Railway. This photograph shows a KP construction camp on the High Plains east of Denver.

DAVID S. DIGERNESS COLLECTION

THIS SKETCH is one artist's rendition of the first railway service to reach Denver. The Denver Pacific Railway and the Kansas Pacific Railroad both arrived at Denver City in the year 1870, within two months of each other. This was several years prior to the development of the concept for a union depot in Denver.

associates on behalf of the Kansas and Denver Pacific railroads and Denver, and by Loveland and his associates on behalf of the UP and Golden, had created great bitterness and soon developed feelings of personal hostility between the rival leaders.

In 1871, Gilpin County voted $300,000 in bonds to aid the Colorado Central Railroad on the condition that it should be built to Central City and that it should be completed to Black Hawk by May 1, 1872. The physical conditions in Clear Creek Canyon prevented the railroad from reaching Black Hawk on time and it lost the bonds. A new arrangement was then made with Gilpin County under which $250,000 of bonds were voted, upon condition that the road be completed from where it was in Clear Creek Canyon to the Forks of the Creek by September 1, 1872; to Black Hawk by January 1, 1873; and to Central City within one year. The road was finished to the Forks of the Creek Station on time, to Black Hawk on December 15, 1872, and to Floyd Hill — on the Georgetown division of the narrow-gauge — in March of 1873. Progress beyond Black Hawk at that time was interrupted by engineering difficulties; however, before a satisfactory solution was found, the Panic of 1873 brought all railroad construction in the Clear Creek valleys and canyons to a standstill, and it remained stalled for several years; nevertheless, the finished portion of the line was in full operation with stages and wagons making connections at both terminals.

The narrow-gauge line from Golden up Clear Creek Canyon was a tedious piece of construction and a remarkable example of a most difficult railroad project. Beyond Golden, Jesse found the track to consist of a succession of curves as it twisted its way up the gorge, crossing back and forth from one side to the other, following the creek that flows through this great, contracted, water-worn rock chasm of magnificent beauty.

The influence of Clear Creek Canyon on the ordinary tourist, perhaps, will be comparatively transient, fading into a dreamlike memory of amazing mental impressions. Not so with the man who dwelt untutored for many years amid these stupendous hills and abysmal gorges. His imagination, once aroused and enlarged, continued to expand; his fiction, once created, hardened into fact; his veracity, once elongated, stretched on and on forever. Of all natural curiosities, he was the most curious — more marvelous than even Clear Creek Canyon itself. This amazing man — who helped push the railroad through this canyon — was Captain E. L. Berthoud.

Another part of the Colorado Central's plans to isolate Denver and concentrate railroad influences at Golden, was the construction of a road south from Golden to a junction with the D&RG line near Littleton. By this time, the Rio Grande was pushing south beyond Colorado Springs. The roadway of about 18 miles was graded in the spring and summer of 1873; however, like the Golden-to-Julesburg enterprise, the Golden-to-Littleton project suddenly went down in the financial gale of that year and was never revived.

The years 1872 and 1873 were times of great turmoil between rival railroads, rival projects and rival builders in the Denver region. Soon after the Kansas Pacific and Denver Pacific roads had been completed — thus forming the "Eastern Division" of the Union Pacific, as contemplated by Congress — intense animosity broke out between the two routes. Congress intended that the UP and KP should cooperate, exchanging and dividing business with each other. To better enable the Kansas Pacific to meet its adversary, John Evans resigned the presidency of the Denver Pacific early in the month of March, 1872, and was succeeded by General R. E. Carr, who also was president of the Kansas Pacific. This gave the latter absolute control of the route from Kansas City to Cheyenne, and made the Eastern Division one route in fact, although the Denver Pacific remained nominally a separate corporation.

58

Chapter V

Denver Union Depot

DURING NOVEMBER OF 1879, incorporation papers were drawn for the Union Depot & Railroad Company in Denver, with a proposed capitalization of $400,000. W. S. Cheesman was elected president by representatives of the five railway companies backing the venture (acting as trustees).

The site chosen for the depot and grounds was between Sixteenth and Eighteenth streets, embracing 12 acres. The front of the proposed structure would face existing trackage of the Denver Pacific and the Denver & Rio Grande.

By the time construction got underway on the depot in 1880, the railway companies owning the terminal company were the Colorado Central; Denver, South Park and Pacific; Denver & Rio Grande; and the Union Pacific, which had absorbed two of the original incorporators: the Denver Pacific and the Kansas Pacific, and thus had obtained twice the number of shares in the terminal company owned by any of the other roads.

The D&RG and the DSP&P got the job of hauling building stone to the construction site. The D&RG laid track into a stone quarry southeast of Castle Rock and proceeded to move rock to Denver at a fairly rapid pace. This material was pinkish-gray rhyolite. The rhyolite was accented with pink sandstone hauled in from Morrison over the DSP&P.

The center of the depot was dominated with an ornate tower some 200 feet tall and 32 feet square at the base. The main structure itself was 504 feet long, consisting of a center section 158 feet long and two end sections, each 83 feet in length, both of which were connected to the center section by what might be called intermediate extensions. The endfaces were 65 feet wide and were duplicates of half the center section. The entire building was two stories tall, covered with four-slope roofs. When the depot was opened to the public in 1881, there was no celebration or fanfare of any kind.

By the end of the 1880's, the Denver Union Depot was serving nearly 100 trains a day. By 1892, traffic had grown to such an extent that single-story extension sheds had to be added to each end, making the structure some 880 feet in length. Two years later, this impressive structure lay in smoldering ruins, the victim of an unfortunate electrical fire. On March 18, 1894, an electric wire short-circuited beneath added to each end, making the structure some 880 feet in length. Two years later, this impressive structure lay in smoldering ruins, the victim of an unfortunate electrical fire. On March 18, 1894, an electric wire short-circuited beneath the roof, starting a fire that rapidly spread to the entire center section of the depot. Before firemen could bring the inferno under control, the tower and most of the center section had been burned out.

Reconstruction began almost immediately, and the rebuilt structure generally was considered to be an improvement over the original building. The clerestories were eliminated from the upper roof, since the new structure was being rebuilt with a roof having a less-inclined slope. The new tower was of a much better design, featuring a four-faced clock, which gave the building more the look of depots in turn-of-the-century America. Although the new design might have been thought of as being less distinctive, it must be admitted that the original tower had somewhat of a "churchy" appearance.

Denver Union Depot gained a new railway organization as a tenant in 1898 — the Colorado & Southern Railway, basically a combination of the Old South Park narrow-gauge lines and the former Denver & New Orleans (UPD&G) standard-gauge lines. In the process, the C&S obtained a one-fifth ownership of the union terminal.

As a result of the creation of the Colorado & Southern, a terminal reorganization took place; the new company was named the Union Depot & Railway Company. The C&S and the D&RG each owned a one-fifth interest, while the UP owned three-fifths interest. Other railway lines using the terminal paid rent, either to the terminal company or to the railway performing services for them.

In 1912, the terminal company again was reorganized. This time, all six railway companies using the terminal were involved in the reincorporation; these companies were the AT&SF, CB&Q, CRI&P, C&S, D&RG, and the UP. The new corporate name was Denver Union Terminal Railway (DUT), the name the company retains as of this writing.

During the DUT reorganization, the depot once again was rebuilt. The center section — tower and all — was replaced with a gray granite structure of massive proportions, featuring columned portals, arched windows and a high waiting-room

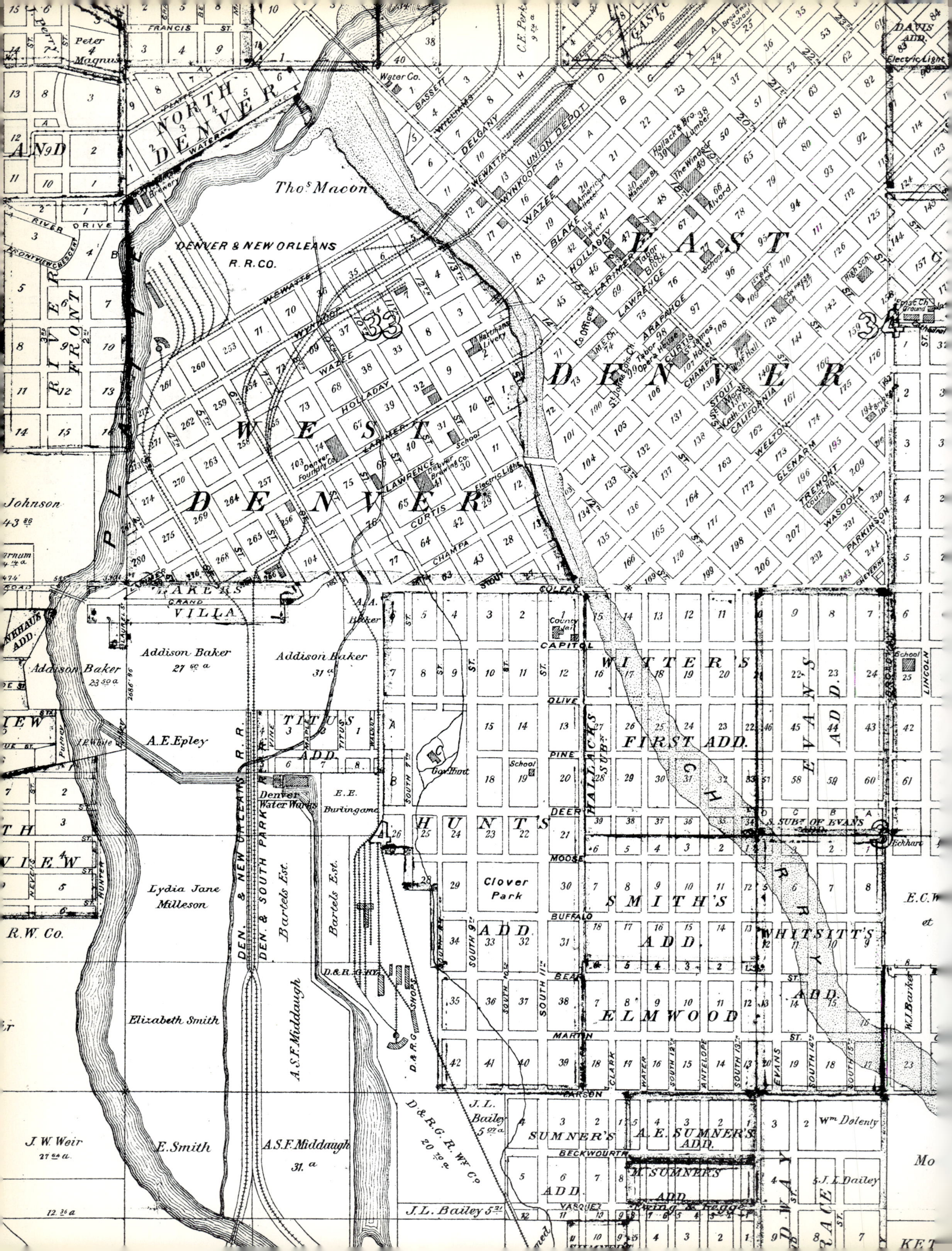

NORTH DENVER
EAST DENVER
WEST DENVER
Thos Macon
DENVER & NEW ORLEANS R.R.CO.
UNION DEPOT
WITTER'S FIRST ADD.
HALLACK'S SUB.
HUNT'S ADD.
SMITH'S ADD.
ELMWOOD
WHITSITT'S ADD.
EVANS ADD.
A.E. SUMNER'S ADD.
M. SUMNER'S ADD.
SUMNER'S ADD.
TITUS ADD.
GRAND VILLA
Clover Park
Addison Baker
A.E. Epley
Denver Water Works
E.E. Burlingame
Bartels Est.
Lydia Jane Milleson
Elizabeth Smith
A.S.F. Middaugh
E. Smith
J.W. Weir
J.L. Bailey
D.&R.G. R.Wy. Co
D.&R.G. SHOPS
DEN. & NEW ORLEANS R.R.
DEN. & SOUTH PARK
Wm Dolenty
J.L. Dailey
COLFAX
CAPITOL
OLIVE
PINE
DEER
MOOSE
BUFFALO
BEAR
MARTIN
PARSON
BECKWOURTH
CHERRY
PLATTE
RIVER FRONT
Johnson
R.W. Co.

ceiling. At the same time, the station trackage was raised some five feet to allow for underground access to island platforms, which were provided with umbrella roofs. Also included in the reorganization was an official terminology change for the name of the physical plant, from "union depot" to "union station," although the building itself continued to be referred to as a "depot."

Today, Denver Union Station retains the look it assumed in 1912, even though our country's transportation policy — or lack of policy — has permitted a great decline to occur in railway passenger-train operations. Instead of encouraging efficient railway passenger-train usage, our Federal government has placed nearly all its emphasis on forms of transportation that are not only wasteful of fuel, but often are gross polluters of the atmosphere. Beyond that, it is shocking that the U. S. government has not worked for the electrification of all primary mainline railways in the United States — as other major industrial countries have done. Washington certainly must be aware that electric locomotives can receive their power from almost any source of energy — from coal-fired and hydroelectric plants to the very latest in geothermal and solar-energy plants.

It remains to be seen what will happen to Denver Union Station. Will this maginificent structure go the way of so many other elegant Denver buildings, such as the Windsor Hotel and the Tabor Opera House? Or will common sense at last prevail?

DAVID S. DIGERNESS COLLECTION

ADDITIONAL TRACKAGE had been laid in front of Denver Union Depot by 1883.

DAVID S. DIGERNESS COLLECTION

THE NEWLY COMPLETED Denver Union Depot was opened to the public on June 1, 1881. At the time, the original track ran in front of the building. The building measured 65x503 feet and the tower was 180 feet tall. Note the "harp" switchstands.

DAVID S. DIGERNESS COLLECTION

THE GROUNDS in front of Denver Union Depot had been landscaped with trees and grass by 1885, considerably improving the overall appearance of the building. In the process, the station trackage was moved to the west side of the structure.

OPPOSITE: An early map of Denver shows Union Depot in the top center, and the Old South Park terminal location at the Denver & New Orleans Railroad site, in the top left.

DAVID S. DIGERNESS PHOTO

DENVER UNION STATION during the steam era was a busy place both day and night. Here, a Burlington steamer was recorded in a time exposure by the author. Alas! The sights, sounds and smells of Union Station no longer eminate from that romantic form of motive power, the steam locomotive!

THIRD-RAIL TRACKAGE was very much in use at Denver Union Depot during the early days — as this late-1890 view portrays. Narrow-gauge passenger trackage continued to be in evidence in Denver until the Colorado & Southern gave up service to Leadville in 1937.

DAVID S. DIGERNESS COLLECTION

W. H. JACKSON PHOTO
RICHARD RONZIO COLLECTION

A UNION PACIFIC fast-passenger train was awaiting departure at Denver Union Depot, circa 1885, shortly after the tracks were placed on the west side of the building.

DAVID S. DIGERNESS COLLECTION

LOCOMOTIVE NUMBER 2 of the Colorado & Southern was photographed at Denver Union Depot in 1900 with a passenger train in tow. This engine — named "Hill Top" — was first numbered 31 when built in 1882, then was renumbered 158 in 1885.

DAVID S. DIGERNESS COLLECTION

RUSSELL PHOTO
DAVID S. DIGERNESS COLLECTION

LOCOMOTIVE NUMBER 44, the ''Lake City,'' was ready for departure from Denver Union Depot one sunny day in 1887. This Denver, South Park & Pacific train was headed for the end-of-track near Webster, with combination baggage-chaircar 6 in tow. Formerly numbered ''8,'' this Mason bogie was built in 1879.

DENVER UNION DEPOT was re-constructed in 1895. By this time, electric-trolley tracks had arrived on the scene, and Denver's famous ''welcome arch'' over Seventeenth Street had been erected. The first overhead electric streetcar lines were installed in the downtown section of Denver in the 1890's, converting the extensive cable-car lines to the more modern system.

DAVID S. DIGERNESS COLLECTION

OPPOSITE: An artist's rendition of Denver Union Depot landscaping.

DAVID S. DIGERNESS COLLECTION

COLORADO & SOUTHERN locomotive 5 was ready for departure from Denver Union Depot in 1900. This was nearly the last year the railroads were allowed to use link-and-pin couplers, which were considered to be dangerous. (The changeover to automatic knuckle-couplers took place in 1903.) This handsome little narrow-gauge 2-6-0, with 40-inch drivers, was constructed for passenger service in 1884.

C. A. YONT PHOTO
DAVID S. DIGERNESS COLLECTION

ON A COLD WINTER DAY in the early 1900's, a south-bound passenger train was waiting for orders at Denver Union Depot.

Denver, South Park & Pacific

RAILROAD.

CONNECTIONS

AT DENVER

With all Trains over the

Union Pacific,
Kansas Pacific,
Denver & Rio Grande,
Atchison, Topeka & Santa Fe,
Colorado Central, and
Denver & Boulder Valley

RAILROADS,

—FOR ALL POINTS—

EAST AND WEST, NORTH AND SOUTH.

L. C. McCLURE PHOTO
DAVID S. DIGERNESS COLLECTION

THIS VIEW of Denver's depot was taken after brick was laid in the street, around 1900. An automobile is parked beside the curb, although horse-drawn vehicles provided the primary mode of transportation at the time.

WITH A HISS from the cylinder cocks, newly renumbered C&S Number 44 starts up in front of Denver Union Station in 1899. The agile gentleman in front of the engine is demonstrating the typical ''boomer'' technique of stepping on a moving locomotive pilot. Such risky practices disappeared with the advent of unions and the progressive ''safety first'' campaigns of the Twentieth Century.

Wells Fargo & Company Express.
D&RG
619

A VAST NETWORK of standard- and narrow-gauge tracks served Denver Union Depot as of 1905. Denver & Rio Grande Number 619 [in the left foreground] had been assigned switching chores, which meant that she had been demoted from road service. Normally, an engine such as the 619 would have been assigned to freight service, due to her 2-8-0 wheel arrangement. Thousands of passengers arrived and departed from Denver Union Depot each and every day of the week.

DAVID S. DIGERNESS COLLECTION

W. H. JACKSON PHOTO
DAVID S. DIGERNESS COLLECTION

DENVER UNION DEPOT once had this attractive landscaping on the street side of the building. The depot was the focal point for many of Denver's people, as express, baggage and mail shipments — as well as thousands of travelers — arrived and departed daily. On March 18, 1894, a fire destroyed a large portion of the original

depot. Reconstruction resulted in the structure shown above. A second reconstruction took place in 1914. At that time, the tower was removed and the entire center section was enlarged. In addition — prior to the last rebuilding — extension wings were added at both ends of the depot to handle mail and express.

DAVID S. DIGERNESS PHOTO

IT SHOWS— —the shattering silence that prevails around Denver Union Station in 1977. This was once the scene of shoulder-to-shoulder activity, with travelers constantly pouring on or off trains from all areas of the U.S.A. Today—because of an unbalanced National transportation policy—this terminal is a shadow of its former self.

THE UNION PACIFIC DENVER AND GULF RY. CO.
Frank Trumbull, Receiver

OFFICE OF GENERAL SUPERINTENDENT.

T. F. DUNAWAY,
GENERAL SUPERINTENDENT.

H-27.

Denver, Colorado, April 1st, 1898.

Mr. J. W. Dalbey,
Division Superintendent,
Denver, Colorado.

Dear Sir:-

Below I give you a list of stations on your Division, at which I desire to know the distance of the Post Offices, by the

PORTFOLIO OF OLD DENVER

ALEX MARTIN PHOTO
RICHARD RONZIO COLLECTION

A HORSE-DRAWN STREETCAR rattles along Sixteenth Street in downtown Denver during 1889. The famous Daniels & Fisher clock tower had not yet been built at the corner of Sixteenth and Arapahoe streets when this picture was produced by Alex Martin.

DAVID S. DIGERNESS COLLECTION

THE ARAPAHOE COUNTY COURTHOUSE is featured in the view above, as well as in the views on the following page. Unfortunately, this grand Victorian building—which was located between Court Place and Tremont Street, and extended between Fifteenth and Sixteenth streets—was demolished in 1934 for "reasons" that make little sense today.

W. H. JACKSON PHOTO
RICHARD A. RONZIO COLLECTION

DENVER WAS SERVED by the Arapahoe County Courthouse prior to the creation of the City and County of Denver. The view above shows that the Front Range could be seen from the downtown streets due to the wisdom of early city planners. W. H. Jackson probably obtained this full-color view from the dome of the State Capitol—perhaps even before the structure was completed. The view below shows the courthouse from street level shortly after its construction. At a later date, the roof line was changed to accomodate an additional floor. Notice that changes were made to the tower, also.

ROBERT M. DAVIS PHOTO
DAVID S. DIGERNESS COLLECTION

W. H. JACKSON PHOTO
STATE HISTORICAL SOCIETY OF COLORADO

THE TWO VIEWS on this page were photographed between 1883 and 1885 by W. H. Jackson. Above: An overall view of Denver featuring the Governors Guard Hall, an assembly building for a militia company called the Governor's Guards. Below: Joslin's department store in downtown Denver.

W. H. JACKSON PHOTO
STATE HISTORICAL SOCIETY OF COLORADO

L. C. McCLURE PHOTO
DAVID S. DIGERNESS COLLECTION

L. C. McCLURE recorded the view above of the Burlington - Colorado & Southern downtown ticket office, circa 1910. Notice the large scenic prints in the windows to lure passenger traffic.

IN THE VIEW BELOW, the photographer was looking out from the State Capitol toward the Arapahoe County Courthouse and the Daniels & Fisher tower on Fifteenth Street. At this time, Denver was known as the "Queen City of the Plains."

L. C. McCLURE PHOTO
DAVID S. DIGERNESS COLLECTION

DAVID S. DIGERNESS PHOTO

THE COLORADO CAPITOL is located on a hill on the edge of Denver's central business district. Photographed from the Penn apartments, this view of the capitol also shows the Front Range. Denver's first "skyscraper," the old Daniels & Fisher clock tower, is largely hidden behind the Hilton Hotel. Just beyond the D & F tower, Denver Union Station continues to handle passenger trains.

STATE HISTORICAL SOCIETY OF COLORADO

W. H. JACKSON was in the heart of downtown Denver—at Fifteenth and Lawrence streets—when he photographed this cable car. During the 1880's, Denver had an extensive cable-car system, providing transit service to much of the city.

AT THE TURN OF THE CENTURY, the U.S. Mail sported delivery rigs such as the one in the photograph below. At this time, inter-city mail sorting was done aboard fast trains, providing good service all across the country—a far cry from today's poor postal service.

STATE HISTORICAL SOCIETY OF COLORADO

W. G. CHAMBERLAIN PHOTO
DAVID S. DIGERNESS COLLECTION

PANORAMA OF

OPPOSITE: New brick buildings replaced the earlier wooden structures in Denver shortly after the great fire of 1863. Reconstruction—for those who could afford it—started almost immediately in the business district of the city. The buildings in this view were located on the southeast side of Blake Street, near Fifteenth Street.

THE SCENE AT RIGHT is one of W. G. Chamberlain's early-day views of Denver City. It is believed that the bridge shown was the one that once spanned Cherry Creek at Blake Street in the lower downtown section of the city. Auraria is on the west side of Cherry Creek at this point.

THE PANORAMA BELOW portrays Denver as it appeared from Capitol Hill during the late 1880's. It is from a photograph by W. H. Jackson, one of several he took from this vantage point.

W. G. CHAMBERLAIN PHOTO
RICHARD RONZIO COLLECTION

DAVID S. DIGERNESS COLLECTION

STATE HISTORICAL SOCIETY OF COLORADO

ABOVE: The Denver Chamber of Commerce building looked like this about 1890. The top floor housed the organization that later became the State Historical Society of Colorado.

BELOW: An artist's rendition of the Tabor Grand Opera House, circa 1890.

DAVID S. DIGERNESS COLLECTION

DAVID S. DIGERNESS COLLECTION

THE TABOR GRAND OPERA HOUSE, at Sixteenth and Curtis streets, had just been completed in 1881, when this view was recorded. The Tabor Grand—as it was always called—was of Victorian Romanesque design, one of the most majestic structures ever to grace the landscape of the American West. All the great players—from Oscar Wilde to Sarah Bernhardt—performed here during the '80's and '90's. Built by H. A. W. Tabor, the five-story red-brick building was one of the most lavishly decorated playhouses in the country. Like so many of Denver's great Victorian structures, the Tabor Grand was callously destroyed during the post World War II "ticky-tacky" urban-renewal period. On the theatre curtain, Kingsley's lines were inscribed—so descriptive both of Tabor's fortunes and the fate of his creations—"So fleet the works of man, back to earth again, —Ancient and holy things fade like a dream."

THE FOURTH OF JULY, 1890, was celebrated by the laying of the cornerstone of the new State Capitol in Denver. The castle-like tower rising in the background is the foundation for the iron dome, which later topped the building. This dome was capped by Colorado gold leaf, worth some $4,000 in 1896, when this edifice was completed. The native-granite structure has a floor plan in the form of a Greek Cross, while the building as a whole is of neo-Classic design.

STATE HISTORICAL SOCIETY OF COLORADO

STATE HISTORICAL SOCIETY OF COLORADO

THE COLORADO MILITIA was called out by the Governor in 1894 when a dispute between the State and the City of Denver brought about rioting. The Silver Panic of 1893 had brought on the worst financial crisis in Colorado's brief history—very nearly ruining the State's economy and throwing thousands of men out of work. This scene portrays just one of the tumultous events directly or indirectly caused or effected by this severe depression.

THE GENTLEMAN BELOW was a member of one of Denver's riding and driving clubs at the turn of the century. The State Capitol is in the background.

STATE HISTORICAL SOCIETY OF COLORADO

— THE PIONEER MONUMENT —
At Broadway and West Colfax Avenue

DAVID S. DIGERNESS PHOTO

DAVID S. DIGERNESS COLLECTION

WORKERS POSED in front of a Denver brick-making plant before the turn of the century. Typical of the many small enterprises of this type, this plant produced bricks from local fireclay, a non-metallic mineral. Denver's great fire of April 19, 1868, resulted in a loss estimated at $350,000. Brick quickly became popular in new construction work following the fire, and kilns were built throughout the city. This great demand caused the brick industry of Denver to be launched.

THE STAFF OF THE U.S. MINT, on West Colfax Avenue between Cherokee and Delaware streets, posed for their picture many years ago. This is one of two U.S. gold repositories (the other one being at Fort Knox, Kentucky). Coinage of pennies and silver pieces has gone on here since 1906, 2 years after the completion of the present plant.

DAVID S. DIGERNESS COLLECTION

DAVID S. DIGERNESS COLLECTION

A CARRIAGE typical of many owned by the more affluent residents of Denver during the 1880's and 1890's is shown above.

OVERLEAF: The Wall Street of the West—Seventeenth Street—is lined with banks and other financial institutions. The building at the extreme right is the famous Brown Palace Hotel. Denver Union Station is at the far end of the street.

DAVID S. DIGERNESS PHOTO

FIRST NATIONAL
DREXEL

DAVID S. DIGERNESS COLLECTION

DURING THE 1890's Denver's electric power was provided by generating stations such as the one shown on this page. All of these power plants were coal-fired, utilizing the fuel most abundant in the Rocky Mountain West—a fact partially ignored in recent years. Today's electrical engineers would look with disbelief at the lack of safeguards and the primitive wiring in this plant.

DAVID S. DIGERNESS COLLECTION

PAGES 90 -- 95: W. H. Jackson photographed several outstanding panoramas of Denver from Capitol Hill during the late 1880's. Three of them are reproduced

W. H. JACKSON PHOTO
DAVID S. DIGERNESS COLLECTION

herewith. A considerable amount of construction could be seen from this viewpoint—looking toward the Front Range of the Rockies. However, very nearly all of the fine

Victorian structures visible in these views have been torn down and carted, piece by piece, to the city dumps.

Familiar buildings appear as follows: The Arapahoe County Courthouse, pages 91 and 92; the Tivoli brewery

W. H. JACKSON PHOTO
DAVID S. DIGERNESS COLLECTION

on what is today's Auraria college campus, page 93; the Tabor Grand Opera House, page 94; and the tower of

Denver Union Depot, along with the once-famous Windsor Hotel, page 95.

W. H. JACKSON PHOTO
DAVID S. DIGERNESS COLLECTION

STATE HISTORICAL SOCIETY OF COLORADO

THE INTERIOR VIEW above depicts the lavishly decorated library of one of Denver's Victorian mansions. The relatively few "well-to-do" citizens of the period lived very well indeed. Meanwhile, large numbers of those not so fortunate felt lucky to have any kind of roof over their heads—especially after the Silver Panic of 1893 forced thousands of Coloradans into a state of poverty.

THE OLD SOUTH PARK'S DENVER ENGINE TERMINAL

PAUL STRADLEY COLLECTION

NUMBER 537 was the largest locomotive operated by the Colorado & Southern Railway. She was leased to the C&S by the parent CB&Q in 1930 and was not returned until 1939, after the C&S had abandoned most of its narrow-gauge lines. Here, Number 537 is at the coaling tower in the C&S engine terminal—in the very heart of Denver.

OTTO C. PERRY PHOTO
DAVID S. DIGERNESS COLLECTION

NUMBER 72 (above) was built for the Union Pacific, Denver & Gulf in 1897, and was purchased by the Colorado & Southern in 1899. Number 70 (below) was a slightly smaller engine than Number 72, and was built for the Denver, Leadville & Gunnison in 1890.

WILLIAM A. GIBSON PHOTO

PAUL STRADLEY COLLECTION

NUMBER 73 was one of three 2-8-0's built for the UPD&G in 1897, and later transferred to the C&S in 1899. These engines were numbers 71, 72 and 73. Here, the 73 poses in front of the C&S roundhouse in Denver.

THE SCENE BELOW depicts a typical lineup of C&S narrow-gauge engines in Rice Yard. Located near downtown Denver, this dual-gauge terminal was a busy place 'round the clock during the steam era.

LAD G. AREND PHOTO
DAVID S. DIGERNESS COLLECTION

D. E. ROGERS PHOTO

NUMBER 75 was pulling out of the Denver yard with a freight train when D. E. Rogers photographed the view above. This 2-8-0 was a Brooks product of 1898, built for the Colorado & North Western and obtained by the Colorado & Southern in 1921.

OTTO PERRY took this fine portrait of C&S Number 68 in Denver during 1931. Built in 1890, this Baldwin engine originally was Number 271 on the DL&G.

OTTO C. PERRY PHOTO
DAVID S. DIGERNESS COLLECTION

DAVID S. DIGERNESS COLLECTION

THE TWO ENGINES on this page portray former DSP&P 2-6-0's. Number 8 was the 71 on the DSP&P, while Number 4 was the 39. Both were built by the Cooke works in 1884; Number 8 served the road until 1938, but Number 4 lasted only until 1934. Engines 4 to 10 were classified as passenger-train motive power and all were basically alike. This series of 2-6-0's was the pride and joy of the Old South Park for many years.

OTTO C. PERRY PHOTO
DAVID S. DIGERNESS COLLECTION

PAUL STRADLEY COLLECTION

DICK KINDIG recorded this view of Number 68 in the Denver terminal during 1937. This Baldwin 2-8-0 was built in 1890 for the Denver, Leadville & Gunnison as their Number 271. She lasted until the C&S narrow-gauge mainline was abandoned in 1938.

C&S NUMBER 37 was at the Denver roundhouse in this photo of 1917. She was originally numbered 41 for the DSP&P when built in 1883 by Cooke. Her number in 1885 was 198.

OTTO C. PERRY PHOTO
DAVID S. DIGERNESS COLLECTION

DAVID S. DIGERNESS COLLECTION

DENVER, LEADVILLE & GUNNISON locomotive Number 57 was photographed in Denver around 1900. She was named the "Buena Vista" and was numbered 24 when new in 1880.

THE TRIM LINES of the "Ruby"—a Mason bogie—were photographed in the 1880's. The tall Congdon stack was an unusual early design to catch red-hot cinders—while the kerosene headlamp was typical of the period. This handsome little engine was built in 1879, and originally came with fine-scroll lettering and lining.

DAVID S. DIGERNESS COLLECTION

OTTO C. PERRY PHOTO
DAVID S. DIGERNESS COLLECTION

LOCOMOTIVE NUMBER 21 was under steam when photographed in 1916 in Denver. She was numbered 29 when new in 1882, and was renumbered 156 in 1885.

NUMBER 9, A C&S 2-6-0 was at the coal chute in Denver, August 27, 1937. Originally DSP&P Number 72, and renumbered 114 in 1885, she is now on display at Hill City, South Dakota.

RICHARD H. KINDIG PHOTO
DAVID S. DIGERNESS COLLECTION

DAVID S. DIGERNESS COLLECTION

COLORADO & SOUTHERN locomotive Number 6 was eased out of the Denver roundhouse in 1937. This was formerly the 69 of the DSP&P, and became Number 111 in 1885.

NUMBER 6 served the C&S until the narrow-gauge mainline was scrapped. Like most other C&S slim-gauge locomotives, she saw service on both the Colorado Central and the South Park lines from the beginning of C&S operations.

D. E. ROGERS PHOTO

RICHARD H. KINDIG PHOTO
DAVID S. DIGERNESS COLLECTION

C&S NUMBER 70 was taking a hot water bath beside her big sister, CB&Q Number 6309, on June 19, 1937, in Denver.

DSP&P LOCOMOTIVE NUMBER 15, the "Breckenridge," was resplendent in new paint — a 2-6-6T Mason bogie fresh out of the Mason Machine Works in Taunton, Massachusetts, in 1879.

OPPOSITE, ABOVE and BELOW: C&S numbers 60 and 58 were on one of the turntable lead tracks beside AT&SF engine 3764 on July 14, 1936. The dual-gauge turntable and roundhouse are visible in the distance. DSP&P 2-6-6T Number 55 was photographed at Garrison, Montana, in 1886. She had been leased to the Utah & Northern. She was built in 1879 and originally was numbered DSP&P 22. In 1885 she was renumbered 55.

DAVID S. DIGERNESS COLLECTION

RICHARD H. KINDIG PHOTO
DAVID S. DIGERNESS COLLECTION

DAVID S. DIGERNESS COLLECTION

RICHARD H. KINDIG PHOTO
DAVID S. DIGERNESS COLLECTION

RUNNING EXTRA to Denver, October 15, 1938.

Chapter VI

Denver to Webster

THE SAGA OF BUILDING the "Old South Park" rail lines, and the history of the areas they served, is a dramatic story; and the passage of time, in turn, has brought about some of the most cherished railroad history ever achieved. The Denver, South Park & Pacific Railroad, without a doubt, was one of the most famous of American narrow-gauge railroads. Truly a Denver railroad, it was built with Denver money, by Denver men, for Denver. As was the case with the standard-gauge Denver Pacific Railway, John Evans also was president of the DSP&P at the time of its completion. Now, 100 years have passed since the founding of this iron trail; however, time has not dimmed the brilliance of the record achieved by its builder, Governor John Evans.

Even though little remains of the far-famed "Old South Park," Jesse S. Randall recorded in his supreme records the fabulous story of some of its history — and this goes all the way back to when he was working on Georgetown's *Colorado Miner*. The first mention of any note was a railroad survey that was accomplished by The Denver, South Park & Rio Grande Railroad, organized in February of 1868 by Governor A. C. Hunt. The route of this survey was later used by the DSP&P as far as the Arkansas River. It was run from Denver up the South Platte River through Platte Canyon, crossed over Kenosha Divide into South Park, then went southwest across the park to a low point in the mountains, flanking the park's southwest side, and dropped down into the Arkansas River valley. From there, it ran directly south to the Rio Grande Valley and Santa Fe, New Mexico. It was in this survey that the Platte Canyon was classed as impassable. Shortly thereafter, Governor Evans interested himself in the great natural resources southwest of Denver and viewed them first hand. This included the vast forested areas from Bear Creek to Platte Canyon and west to Kenosha Pass, as well as beautiful South Park and its well-known mineralized areas. He visualized great agricultural areas in lower Bear Creek Valley between Denver and the mouth of Bear Creek Canyon, logging camps to harvest lumber, quarries for mining of dimension rock or building-stone deposits located in the hogbacks at the foot of the mountains. However, greatest of all was his vision of a railroad west, crossing the mountains to the Pacific.

Georgetown and Clear Creek County proposed a bond issue in favor of helping finance the construction of the newly formed Denver, Georgetown & Utah Railway, feeling the Colorado Central Railroad had let them down after it had failed to build into the county in a specified time, as demanded by a bond issue voted to them earlier. The Denver, Georgetown & Utah Railway came about by the reorganizations of three predecessor companies. First of these was the Denver & Georgetown Railroad & Telegraph Company, incorporated February 25, 1868, and in turn, reorganized into the Denver Central & Georgetown Railway Company, which was incorporated November 25, 1868. Two years later, this one was reorganized and named the Denver Central & Georgetown Railroad Company; then, this company, in turn, was reorganized into the Denver, Georgetown & Utah Railway Company, March 9, 1872, with R. E. Carr, president; John D. Perry, vice-president; R. R. McCormick, secretary; and David H. Moffat, treasurer; with John Evans acting as adviser and forwarder of the enterprise.

Projected plans were to build a narrow-gauge railroad from the Kansas Pacific Railroad at Denver City in Arapahoe County, southwest to Bear Creek Canyon, Mount Vernon Canyon, thence via the "high line" route in Jefferson County over the hill. From there, it would drop down to Clear Creek Canyon, east of Idaho Springs, and proceed up the canyon to Georgetown in Clear Creek County, thence westward to Utah over, or under, the Continental Divide. Two branch roads were to connect two very important towns in Gilpin County, one to Black Hawk and the other to Central City. The last reorganization was in part brought about by Georgetown's demands for a railroad. It also was helped by Denver City wanting the mining business from the mountains and its readiness to back a railroad enterprise that would accomplish this goal. On April 15, 1872, Georgetown and Clear Creek County voted $200,000 in financial help for the Denver, Georgetown & Utah Railway, to be used only for its construction within Clear Creek County — and with their own interests in mind — protective clauses and time limits were attached.

In November of that year, Arapahoe County — of which Denver City was the county seat at that time — voted an equal amount to the enterprise. The movement on the part of Kansas Pacific interests was to build a road from Denver

JOHN W. MAXWELL PHOTO

COLORADO & SOUTHERN narrow-gauge 2-8-0 Number 70 churned her 37-inch drivers through the weeds as she headed southbound in 1941. The train was hauling cars billed for delivery to Waterton, South Platte and the Silica Branch. This picture was taken in the Ruby Hill district, on the outskirts of metropolitan Denver.

directly west across the mountains to a connection with the Central Pacific Railroad in Utah, so that their Eastern Division might later become independent of the Union Pacific. This movement aroused the Colorado Central people to extreme activity in Clear Creek Canyon, and was largely the cause of their road being completed to Forks Creek and Black Hawk in the autumn of 1872.

It was hoped by the Honorable John Evans and his associates that joint rights on the Colorado Central Railroad could be worked out with W. A. H. Loveland — mainly on Loveland's narrow-gauge line up Clear Creek Canyon, from Golden in Jefferson County. Evans also hoped to persuade him to allow the addition of a third rail to accommodate narrow-gauge traffic between Denver and Golden.

Unsuccessful efforts to work out these desires formulated the foundation-laying for the later interests that Evans developed. Georgetown was irritated by no evidence of visible construction, and Georgetown's *Colorado Miner* printed this question: "Gentlemen — when shall ground be broken in Clear Creek County?" Later, Georgetown again voted a bond issue to the Colorado Central in their efforts to get a railroad.

By this time, John Evans was devoted to running a railroad into South Park to tap that area's great natural resources. Evans promoted his ideas and planned to build a branch to South Park from the Denver, Georgetown & Utah Railway. No doubt, this was what brought about the organization which was incorporated on October 1, 1872, as the Denver, South Park & Pacific Railway, with John Evans, president; David H. Moffat, Jr., vice-president; George W. Kassler, secretary; Charles B. Kountze, treasurer; and Bela M. Hughes, Joseph E. Bates, Frederick A. Clark, F. Z. Baloman, Henry Crow and Walter S. Cheesman, directors. Projected plans called for the line to be built westward from Denver into the mountains in conjunction with the Denver, Georgetown & Utah, with the newly incorporated railway veering off from the DG&U at some point for South Park, thereby reaching the great natural resources and mining camps in the South Park country.

Evans built a toll road into Bear Creek Canyon to provide access for logging of railroad ties. The Morrison Stone, Lime & Townsite Company was organized October 24, 1872, by the same organizers of the Denver, South Park & Pacific Railway, along with George M. Morrison.

JOHN W. MAXWELL PHOTO

NUMBER 70, of the Colorado & Southern Railway, was photographed in the southwestern part of Denver on July 8, 1941 — south of Bowles Avenue, along the old Platte Canyon Road. This oil-burning 2-8-0 was hauling at least 15 boxcars and a caboose when she clattered past Milepost 11.

W. A. H. Loveland's Colorado Central narrow-gauge line was built to Black Hawk in early December of 1882 with the help of bonds voted by the people of Georgetown. The bonds voted earlier, in favor of the CC's rival — the Denver, Georgetown & Utah Railway — had expired with no sign of construction in Clear Creek County. The promoters of the DG&U, who had almost lost interest in their plans, were further discouraged by Loveland's refusal to grant joint usage rights through Clear Creek Canyon and his refusal to allow the addition of a third rail from Denver to Golden. Loveland was concerned only with the interests of the town of Golden, and had all intentions of closing Denver out of the railroad business from the mountains, thereby making Golden City the railroad center. He planned for his railroad to have exclusive control of the gateway to the heart of the Mineral Belt, mainly by way of Clear Creek Canyon; he had already penetrated into part of the Clear Creek valleys.

John Evans reorganized his DSP&P into the Denver, South Park & Pacific *Railroad*, instead of *Railway*, and this company was incorporated June 16, 1873, with projected plans to construct a narrow-gauge railroad from Denver to and through Platte Canyon, to and across South Park, to connect with shipping on the Pacific Coast by another route; and, in turn, the line was to serve the southern portion of the famed Mineral Belt. Also, a line was projected to connect with Morrison and the Denver, South Park & Pacific Railroad's interests in the nearby stone quarry. The Denver, South Park & Pacific Railroad was so tied-in with the Denver, Georgetown & Utah Railway that the latter was simply absorbed by the Denver, South Park & Pacific Railroad.

The Kansas Pacific secured complete control of the Denver Pacific at this time. This suited Evans, who was all out for Denver. This also caused the Union Pacific to throw their support to Loveland's Colorado Central. Loveland's intentions were to divert all mountain business through Golden and thence to the northeast, thus by-passing Denver (with the backing of the Union Pacific). The great railroad fights of rivalry for tapping The Mineral Belt caused the organization of many different railroad companies by Loveland in attempts to block the plans of Evans.

W. A. H. Loveland and associates even incorporated — on January 18, 1872 — a railroad to by-pass Denver to the south, the Golden City & South Platte Railway Company, with plans to construct from Golden south, along the east side of the hogbacks to Bear Creek, thence southeast to a connection with William J. Palmer's Denver & Rio Grande near Littleton. Then, on July 24, 1873, the Turkey Creek Railway Company was formed, to be built at a point from the Golden City &

A CORRECT MAP OF THE RAILROAD SYSTEM OF COLORADO,

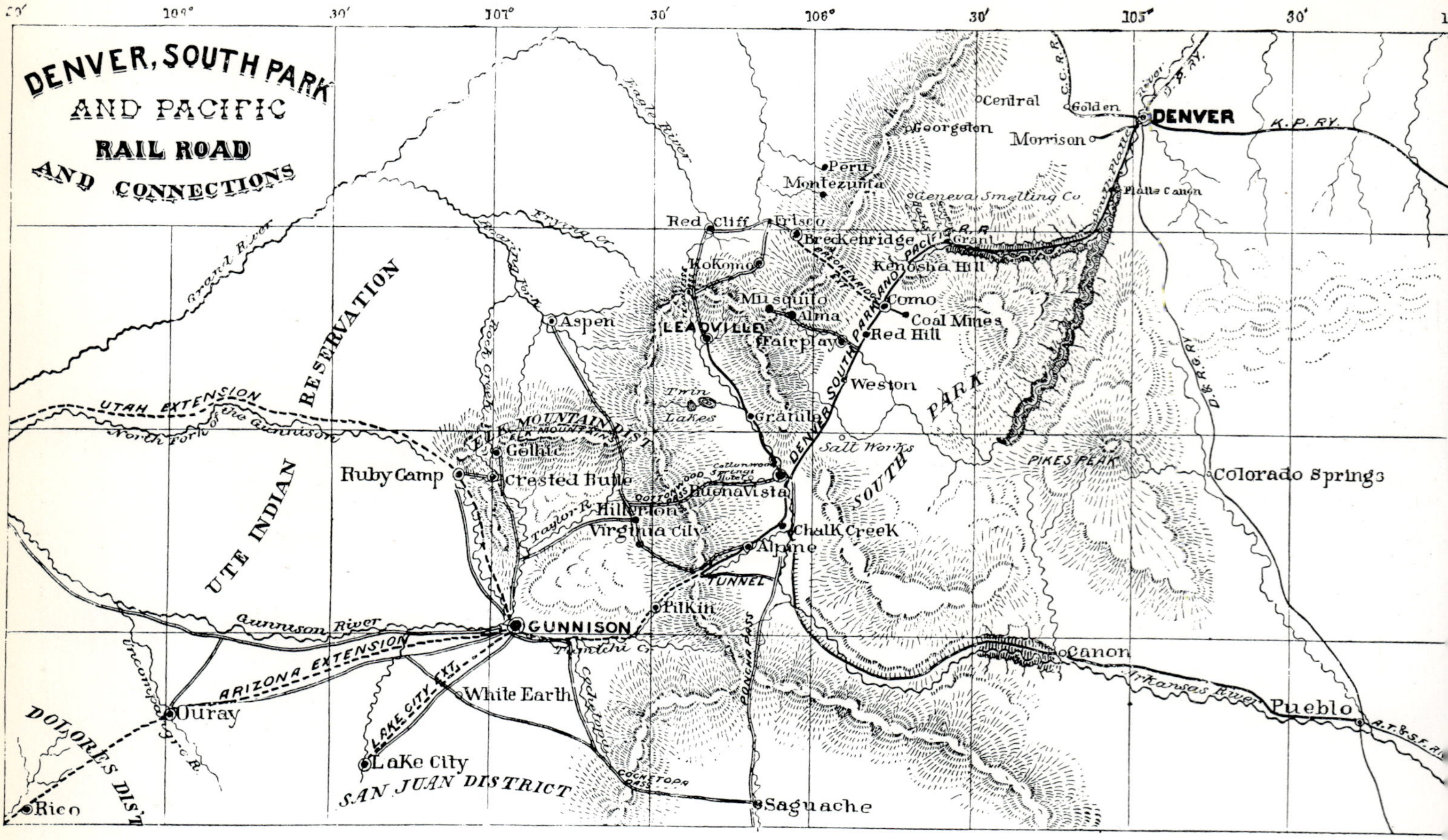

Showing the Location & Connections of the DENVER, SOUTH PARK & PACIFIC R. R.

DAVID S. DIGERNESS COLLECTION

South Platte, close to the confluence of Turkey Creek with Bear Creek. The road then would follow the alignment of Turkey Creek into the mountains via Turkey Creek Canyon, with its destination: South Park.

General William J. Palmer endorsed Loveland's plans for the Golden City & South Platte Railway, which was also backed by the Union Pacific. This was of great concern to Evans and his Denver associates.

Loveland was all out for Golden City, and Palmer was out for his town, Colorado Springs. Evans — on the other hand — was out for Denver and managed to get his ideas across to the Denver voters. Governor Evans went so far as to back his plans with a personal guarantee that the Denver, South Park & Pacific Railroad would be pushed to completion, and Denver would gain access to the mountains and not be by-passed as Loveland, Palmer and the Union Pacific were planning.

On July 28, 1873, Denver voted to transfer the $200,000 voted to the Denver, Georgetown & Utah Railroad to the Denver, South Park & Pacific Railroad, with an added $100,000 being voted at the same time for a total amount of $300,000. Following this, construction was started in August, 1873, and, within a few months, 16 miles of track were laid between Denver and Morrison, where McClellan & Spotswood's Denver & South Park Stage Road entered the mountains. The Denver Railway Association, controlled by the backers of the DSP&P, was the construction company for this earliest construction — with a number of contracting gangs put to work on road grading at different locations.

The first feat was the opening of a narrow-gauge line to Morrison in June, 1874. Road grading along what was to be its mainline proceeded at a slow pace along the South Platte River, southwest to the mouth of the Platte Canyon, and on into the canyon. The financial panic of 1873 — which caused the silver dollar to be dropped from the lists of coins authorized for currency in 1874 — caught up with the DSP&P as it did with so many railroads throughout the country, and temporarily halted further construction. The Denver Railway & Enterprise Company was organized to take over the original construction contract in May, 1875, with little progress to its credit. This was followed by another construction company under the name of the Denver & San Juan Construction Company, on July 6, 1876, also controlled by the backers of the DSP&P Railroad.

Construction was resumed the following October, from Bear Creek Junction southwest to Platte Canyon, where the work of building a roadway up through this natural gateway was pushed with great success. Large forces of men were put to

STAGE LINE

Denver & South Park Stage Company

Daily Line of Four and Six Horse Concord Coaches,

Carrying the United States Mail and Express. Connecting with the Denver and South Park railroad at Morrison, for Fairplay, Alma and all points in Park, Summit and Lake counties. The only Direct Route from Denver to Fairplay, Leadville, Oro City, and all points in California Gulch and the southwestern mines. Train connecting with the coach leaves foot of Larimer street daily at 7 a. m., and connects with coach at Morrison at 8:10 a. m. Through to Fairplay in ONE DAY.

W. C. McClellan & R. J. Spotswood, Proprietors.

DAVID S. DIGERNESS COLLECTION

work, and the grade was completed as far as Milepost 26 in June of 1877. Blasting and building of the grade moved along at a fair rate of speed during the next several months.

Trip to Track-head Near Bailey's

The following year, during the month of October in 1878, Jesse took a first-hand look at the progress of the Denver, South Park & Pacific Railroad by riding to what was called *end-of-track*, also known as the *track-head.* He recorded that the road was completed and in operation as far as Bailey's Ranch on the North Fork of the South Platte River and construction gangs had started to push the grading of the roadbed toward Fairville. The leading officials of the DSP&P at this time were John Evans, president; Walter Cheesman, vice-president; C. W. Fisher, general superintendent; Charles Wheeler, auditor; A. S. Hughes, general freight-and-passenger agent; and L. C. Eicholtz, the road's chief engineer.

The trip started in Denver after Jesse had spent a couple of days in the city on business, following an earlier ride on the Colorado Central Railroad from Georgetown to Denver — a trip that will be brought out in Volume III in relating his records of the Colorado Central. As he rode the South Park's morning train, he descriptively noted the area along the right-of-way and the stations at which they stopped, including their milepost numbers. At Denver, the DSP&P had built an excellent nine-stall roundhouse and had one steam engine inside at the time; outside, there were some boxcars; also, there was a turntable and a water tank. The roundhouse was of rock construction, the rock quarried from Morrison. Departing Denver from the DSP&P's station, located at Sixth and Larimer streets, the train was made up of a number of freight cars and two passenger coaches as it rolled south. It was during this part of the trip that Jesse met Frank Bartlett, the road's surveyor, who was riding on the same train. It was this construction worker who helped provide Jesse with much of the technical information he secured regarding the building of the line through Platte Canyon. It will be recalled that this is the canyon that an earlier survey had erroneously declared "impenetrable."

After passing West Denver at Milepost 2, beautiful hayfields came into view, as well as cultivated farm country. The train came to its first water stop at Bear Creek Junction, the location of a tremendous windmill used to pump water from a well and the creek to the road's large water tank. The milepost was recorded as 7.5, and was the takeoff point for the new route from the Morrison line — the line of Jesse's first recorded trip on the DSP&P. This line was popular because of a resort area and picnic spot at Morrison. Also, the road's business came off this line, mainly with the help of McClellan & Spotswood's Denver & South Park stage lines, and several ore transportation and freight lines using the wagon road from the Alma-Fairplay area to connections with the railroad at the town of Morrison. In addition, there was freight from

JOHN W. MAXWELL PHOTO

THIS NARROW-GAUGE freight train was being pulled by Colorado & Southern Number 70, on its way to South Platte Canyon or Waterton to deliver cars. The locale was west of Littleton, along the old Platte Canyon Road, and the date was July 8, 1941.

Morrison's stone quarries and lumber from the Bear Creek Canyon's logging and sawmill camps.

Along the new line, the train traveled through some very fine ranch country as it approached Littleton, a small nice-looking farm town at Milepost 10.6. After the train left this town, Mr. Bartlett pointed out the following ranches as the train moved along east of the mountains through the Platte Valley: the Brown, the Skelton, the Hussey, the Lilley and the Shallabarger. The clickity-clack of the wheels on the rails slowed, and the train, with a jerk and squeaking of the brake, steamed alongside the depot at Archers at Milepost 16.8. It was while the train was at this station that Jesse was introduced to Isaac W. Chatfield, one of the road's contractors who actually constructed this part of the railroad. The next station recorded was Platte Canyon at Milepost 20.4, where the train once again stopped while the locomotive's tender was filled with water and fuel was taken on. Time permitted Jesse to look over the nice frame depot and inspect a fourstall enginehouse, where one locomotive was being worked on. A wye was used here to turn engines instead of a turntable. New buildings were under construction at the time. The train moved out of this station and entered the canyon with loud blasts of sound from the whistle of the steam locomotive, which echoed in the canyon. Jesse went on to describe the locomotive in the enginehouse. It was Number 3, named *Oro City*, a very unusual type, built by Mason with engine and tender on a common frame, a 2-6-6 tank-type locomotive. This engine was the first of a series of locomotives of this type to be built for the railroad. Then he noted that on the head of their train was a locomotive built by Dawson & Bailey, Number 1, named *Fairplay*, a 2-6-0 with a trailing tender. This was the engine that he looked over in the roundhouse at Denver and he went on to write that it was polished to such a degree that every inch of it shined.

The canyon scenery was especially grand and interesting. Gigantic walls of granite lined each side of the canyon and dwarfed the iron trail. The line ran in the bottom of this spectacular crevice leading into the mountains along the north edge of the river for a distance of three and a half miles. Jesse recorded that the canyon resembled Clear Creek Canyon, through which the Colorado Central Railroad ran, but noted that Platte Canyon seemed to have fewer curves and more straight stretches. Crossing an impressive iron-truss bridge to reach the other side of the river, the train followed the south edge for another three and a half miles, where it crossed back over the river on a second iron trestle. For a total distance of eight and a half miles, the roadbed was carried up to the forks of the river through the narrow, tortuous channel of Platte Canyon, where long stretches of the work were in solid rock. Where fills were impossible, it was necessary to support the line on retaining walls when not blasted out of the hard rock. These retaining walls were most interesting. Mr. Bartlett pointed out that they were dry and very cheap, but very solid, and gave no trouble. They were generally the first work started and were carried along as far as possible before rock excavation was begun.

Mr. Bartlett was proud of his survey, which he worked out personally with the help of his crew. He pointed out that the location of the line was one of his greater challenges. Its construction caused great difficulties, and he stated that the work, in places, was accomplished by men working from ropes suspended from the tops of cliffs to drill holes to blast out the roadbed. In some spots, drill stems were left sticking out of the holes to support platforms from which to work. Fifty-foot high blocks of granite were blasted through to get the grade. The river runs right along the base of the cliffs and allowed no working place, except when the river froze — at which time some of the work was done from the river's icy surface. He pointed out that they proved by their survey that the Platte Canyon line had less curvature and lighter grades than the Colorado Central in Clear Creek Canyon. His surveys were carried out during wintertime, when cold temperatures froze the river hard enough to be walked on, making it possible for him to survey the best location line. Just before the train crossed the second iron trestle — previously mentioned — the train stopped at Deansbury at Milepost 26.8. This station was named for Mr. C. A. Deane, who furnished all the ties for the canyon line. Deane's tie camp and sawmill were located at this point, while his logging operations were carried on both here and up the gulch to the south. Mr. Deane was building a large multi-story brick hotel, and planned for it to be part of a resort area. Mr. Bartlett further pointed out the fact that Mr. Deane delivered to locations down canyon from this tie-cutting camp, close to 30,000 ties, each approximately seven-by-seven inches and seven feet long. These ties were delivered by simply floating them down-river to points where they were lifted out by heavy booms onto the finished grade. Others were hauled upgrade on flatcars. More than 100,000 ties came from this tie camp for the railroad.

JOHN W. MAXWELL PHOTO

JOHN W. MAXWELL PHOTO

COLORADO & SOUTHERN engine 70 had stopped at Bear Creek bridge to take on water when these two color views were recorded on July 8, 1941. The little 2-8-0 was heading south with a trainload of empty boxcars and gondolas. This bridge and water tank were south of the end of Sheridan Boulevard, on the outskirts of Denver.

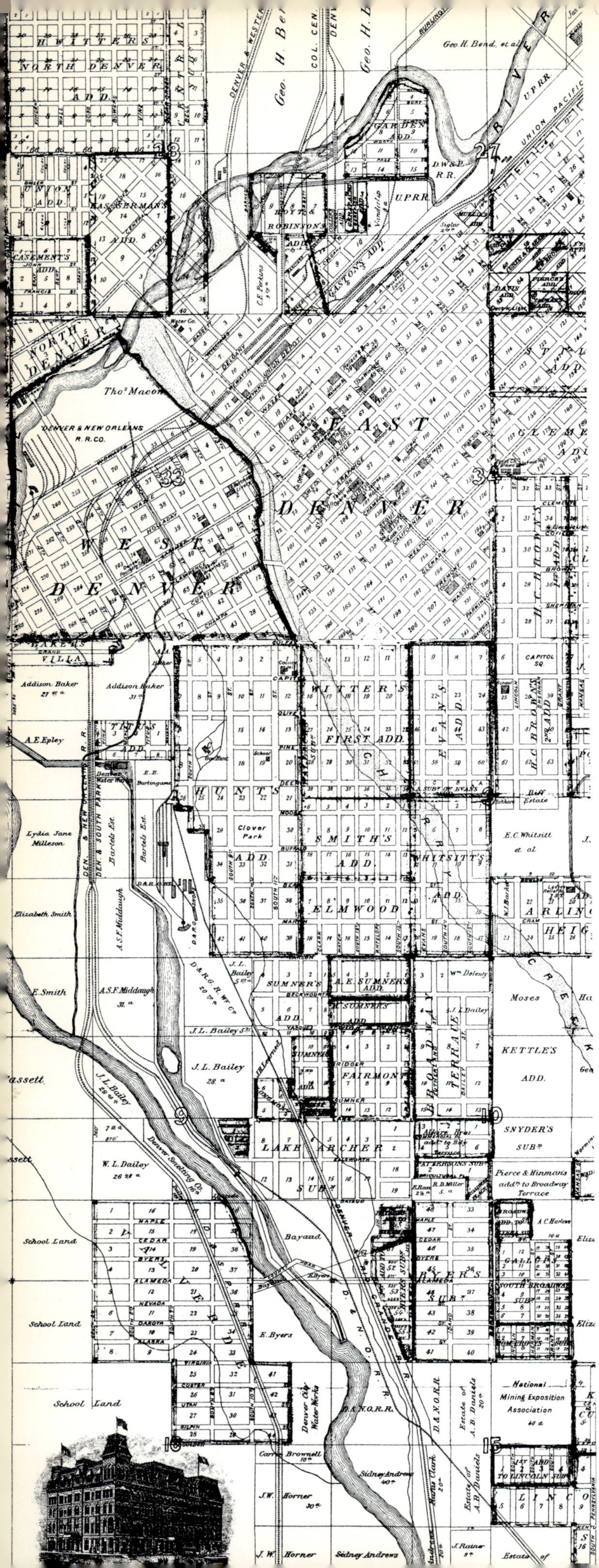

GEORGETOWN COURIER CUT

LEONARD H. EICHOLTZ was the president of the Denver & South Park Railroad Construction Company, and later chief engineer, or surveyor, of the DSP&P Railroad.

Farther along — at the confluence of the South and North Forks of the South Platte River — the train passed South Platte at Milepost 29, where a sawmill was providing much work and a town was forming. At this point, the iron trail turned up the North Fork where the canyon broke out into sloping hills and valleys, more or less thickly covered with timber. Massive weathered, or rounded-rock, outcroppings are features of this area and the following station — called Dome Rock — was named after one of these formations, at Milepost 31.7. Seven miles farther along, the locomotive took on water from Buffalo Tank at Milepost 38.8.

After the large tea-kettle locomotive refreshed herself, the train rolled on up the line. Four more miles brought them to two more stations; the first was Buffalo, at Milepost 39.6, and the second was Pine Grove, at Milepost 42.4, where there was much evidence of activity in the form of erection work for buildings. Continuing up the valley of the North Fork, the train came to Deer Creek at Milepost 48 and another water tank. Whirling off again with the shrill sound of the whistle echoing through the valley, the train was soon within a few miles of the track-head. The beauty of the scenery and the ever-changing view of grandeur that was offered along this railroad was worth anyone's time to ride it, just to see and experience it.

Two more stops — the first at Esterbrook Park, at Milepost 51.6, and then Bailey's Ranch, at Milepost 56.2 — brought the train alongside a newly constructed frame depot. Opposite this station, the Bradford Hill wagon road — from Morrison to South Park — could be seen coming down from the north. The railroad had replaced a portion of the stage line, which used the road that climbed out of the Platte River Valley here, taking off to the northeast over the hill. This wagon road was

OTTO C. PERRY PHOTO
DAVID S. DIGERNESS COLLECTION

WESTBOUND NEAR SHERIDAN JUNCTION during November of 1931, a long train of empties was heading for South Park. There, the stockcars would be loaded with fattened cattle and sheep — coming down from summer grazing pastures. Locomotive 75 was built for the Colorado & North Western (Denver, Boulder & Western) by Brooks in 1898, and was purchased by the Colorado & Southern in 1921. Green flags next to the smokestack indicate that a second section was following this two-engine train.

OTTO C. PERRY PHOTO
DAVID S. DIGERNESS COLLECTION

TAKEN NEAR OVERLAND PARK, Denver, Colorado & Southern locomotive Number 22 was heading for Grant with a trainload of fishermen. This scene, photographed during June of 1918, is within the present city limits of Denver.

RICHARD H. KINDIG PHOTO
DAVID S. DIGERNESS COLLECTION

ARRIVING AT SHERIDAN JUNCTION on October 15, 1938, a trainload from the Silica Branch was heading toward Denver. This "extra," pulled by C&S 2-6-0 Number 9, had 28 cars in tow, and was traveling at the astounding speed of 10 m.p.h.

part of the South Park Stage Road; and, upon the arrival of the railroad at Bailey's Ranch, the Denver & South Park stage line of McClellan & Spotswood once again changed their eastern terminal, utilizing Bailey's Ranch for this purpose. This stage line only recently had its eastern terminal in the town of Morrison, with connections there with the DSP&P's Morrison line. This change had cut out some extremely rough road over the hills that the stage previously had to travel. The stage line's many Concord coaches and their six-horse teams — along with most wagon freighters using the advancing railhead — were concentrating their business on all areas to the west. The railroad handled all the passengers and most of the freight business which was coming over the wagon road. This business originated from South Park and beyond — namely Fairplay, Alma and other locations in Park County, and the Leadville area in Lake County, as well as from the Breckenridge area on the west side of the Great Divide. The South Park stage line connected at end-of-track for points west, with their terminal moving with the track-head.

After a short time at Bailey's, where the engine switched the boxcars onto a siding, the train headed for the track-head, where everyone except Jesse got off the train to travel on west by stagecoach. Jesse noted that a construction train was in front of them at the track-head before stagecoaches came alongside to transfer the passengers.

The principal destination for most passengers was Leadville, by way of Kenosha Divide, into South Park, Fairplay and Weston Pass. However, Jesse returned to Denver via the DSP&P; then took the Colorado Central back home to Georgetown, thus ending his interesting fact-finding trip.

Later on, another trip provided Jesse with additional first-hand observations of the construction progress made on the DSP&P early in May of 1879. He also concerned himself with the fact that Leadville had taken away from Georgetown the title of being the top silver-producing camp. After gathering all the latest information on the railroad and collecting some mineral specimens on a side trip, he traveled all the way through to Leadville to view this bonanza camp. He reported on new depots at three points beyond Bailey's, including descriptive data: Fairville, a small frame structure at Milepost 58; Geneva, a larger frame structure at Milepost 66.2, where ore from Geneva Valley was being loaded into railroad cars; and Webster, at Milepost 69. About a half-mile east of Webster, the steepest grade along the line was encountered — four percent. The station was located at the bottom of a very difficult mountain pass, over which the wagon road went directly to the west. A construction camp was located nearby for work that was in progress on the railroad over Kenosha Divide. The silver stampede to Leadville had caused great activity and throughout the previous winter, hundreds of men had been employed

DAVID S. DIGERNESS COLLECTION

UNION PACIFIC, DENVER & GULF locomotive Number 59, built by Brooks in 1880, was rebuilt in 1885 with larger drivers, measuring 38 inches. She was photographed circa 1895 on a test run out of Morrison — shown here on the Bear Creek bridge headed east. This engine was formerly owned by the Colorado Central, originally numbered 9, and later, 151. She was sold to Little Book Cliff Railroad in 1898.

full-time during all types of weather — working on the steep mountainsides and up rugged ravines, cutting out or building up a roadbed for what was termed an *aerial tramway.*

Jesse took advantage of the eating station at Webster, which cost him $1.00 for a dinner. It is interesting to find out that he was traveling on an "editorial railroad pass" furnished free by the railroad, for which he, in turn, furnished advertising in his newspaper.

The fare for the rail trip at the time was $7.00 from Denver to Webster. Jesse examined the details of the great project still in progress — observing and talking with workers. Trackage was in place more than halfway up the grade to the divide, and he was allowed to ride the work train to observe the progress. The train made a trip up and back, carrying workers on a flatcar. From Webster (known earlier as Halls Valley), the road took a zigzag course, which consisted of many curves enabling it to gain the required elevation without an excessively heavy grade. The line first went in one direction up Halls Valley for some distance, then crossed the South Platte River on a 55-foot wooden trestle — along a large 30-degree lefthand curve — to reach the other side of the valley. In so doing, the line turned back the opposite direction, along the steep and rugged mountainside, leaving the valley floor and the North Fork of the South Platte River. While traversing the hillside high upon the mountain, the route curved to the west, following the north side of the main gulch, gaining elevation necessary to cross the great Kenosha Divide. Along this part of the railroad grade — which was cut out of the side of the mountain — one could look down on Webster far below in the valley; and still further down the hillside (below the grade in the gulch) was the steep wagon road. Beyond this point, after gaining considerable elevation, a feeling of admiration of the engineering skill displayed, came over Jesse as the train proceeded through cuts of projecting mountain spurs and over high fills, crossing side gulches to end-of-track. The roadbed continued to surmount the great Kenosha Divide at a point called Kenosha Pass, where the stage road crosses. The remaining iron was to be laid to the summit by the middle of the month, at an elevation of 10,139 feet above sea level, according to the railroad engineer's survey, or 9,928 feet, according to Professor Hayden's government surveys.

The mining camps using the railroad for transportation of their ore were those of Park, Summit, Lake and Gunnison counties. They hauled their ore to Webster, where stagecoaches also were leaving regularly. The counties served by the railroad included such locations as the great South Park, the areas around the upper tributaries of the South Platte River, and the upper Arkansas River tributaries — mainly California

T. S. LOVERING PHOTO
U. S. GEOLOGICAL SURVEY

THE LANDSCAPE WEST OF DENVER, in 1933, was nearly a wilderness area, except for the small town of Morrison—which was tucked into the ravine along Bear Creek, visible in the left foreground. Above the town of Morrison are the spectacular geologic landmarks known as the Red Rocks. Near the center of this photograph, looking north, the Hogback ridge runs for several miles. The grade of the DSP&P's Garfield Branch ran beside the lower ridge to the right of the Hogback. In the foreground, the grade on the east side of Soda Lake is visible, with the mainline of the South Park heading off to the right for Denver. The Soda Lakes spur, built in 1878 and abandoned in 1919, ran for about one-quarter mile to a small marble quarry. Soda salts were obtained here.

THIS PANORAMIC VIEW of Morrison—as the town looked in the 1890's—shows the Denver, South Park & Pacific's railway facilities and passenger depot in the center of the scene. The Morrison Stone, Lime & Townsite Company was organized on October 24, 1872, by George M. Morrison and the original organizers of the railway. A trainload of rock from Morrison's quarry was about to depart from the station, with a combination baggage-chaircar in tow.

W. G. CHAMBERLAIN PHOTO
DAVID S. DIGERNESS COLLECTION

RICHARD RONZIO COLLECTION

THE TOWN OF MORRISON — pictured here in the 1890's — looked somewhat like a modern-day Western movie set. Locomotive Number 110 of the Denver, Leadville & Gunnison Railroad was pulling a boxcar lettered "Union Pacific," the only boxcar in a train of flatcars loaded with cut stone. Passengers bound for Denver were boarding a combination coach beside the depot.

A YOUNG LADY ON THE PILOT of locomotive 194 was stretching out the white flag, mounted on the pilot, indicating the train to be an extra. One pert, young lady was leaning out the engineer's side of the cab over the lettering designating the Union Pacific ownership of this Denver, Leadville & Gunnison train. On closer examination, the initials of the latter may be seen on the tender and sand dome, below the number. This was the life style of Morrison in the early 1890's, with the arrival of crowds of tourists.

DAVID S. DIGERNESS COLLECTION

DENVER,
SOUTH PARK
And Pacific Railroad.

GOING WEST.			*GOING EAST.*	
NIGHT EXPRESS P. M.	DAY EXPRESS A. M.	STATIONS.	DAY EXPRESS. P. M.	NIGHT EXPRESS A. M.
7 00	7 30	*Lev.* . . . DENVER . . . *Arr.*	8 05	7 00
7 35	8 05	*Arr.* BEAR CREEK Jc . *Lev*	7 35	6 20
8 15	8 45	" . . PLATTE CANON . . "	6 40	5 25
8 45	9 10	" . . . DEANSBURY . . "	6 16	4 55
9 40	9 57	" . . PARK SIDING . . . "	5 33	3 57
10 02	10 12	" . . . . BUFFALO . . . . "	5 16	3 37
10 50	10 51	" . . . . CROSSON'S . . . "	4 35	2 55
11 20	11 15	" . ESTABROOK PARK . "	4 15	2 30
11 37	11 27	" . . . . BAILEY'S . . . "	4 05	2 15
12 01 A. M.	11 42	" . . . . SLAGHT'S . . . . "	3 53	1 57
12 35	12 10 p. m.	" . . . . . GRANT . . . . . "	3 30	1 25
1 50	12 51	" . . . . WEBSTER . . . . "	2 52	1 10
2 40	1 30	" . . . . KENOSHA . . . . "	2 15	12 25
3 00	1 55	" . . . JEFFERSON . . . "	1 55	12 01 p. m.
3 25	2 30	" . . . . . *COMO* . . . . . "	12 55	11 25
3 40	2 50	" . . . . *RED HILL* . . . . "	12 30 a. m.	11 00
4 08	3 16	" . . . . . GARO'S, ... "	11 50	10 24
4 45	3 45	" . . PLATTE RIVER . . "	11 15	10 00
5 30	4 20	" . . . . DIVIDE . . . . "	10 30	9 05
5 55	4 40	" . . . . McGEE'S . . . . "	10 05	8 40
6 30	5 25	" . . . *BUENA VISTA* . . "	9 20	7 45
		" . . . . GRANITE . . . . "		
		" . . . . MALTA . . . . "		
2 00 p.m.	p. m.	*Arr* . . LEADVILLE . . *Lev*		a. m. 10 00
		MORRISON BRANCH,		
p. m. 5 45	a. m. 8 30	*Lev.* . . . DENVER . . . *Arr.*	a. m. 8 00	p. m. 5 00
6 26	9 17	*Arr* . . BEAR CREEK . . *Lev*	7 26	4 26
7 00	10 00	" . . . MORRISON . . .	7 00 a. m.	4 00 p. m.

EXPRESS TRAINS

Leaving DENVER make the following Connections:

At Morrison, with Stage Line for ***Evergreen*** *and* ***Brookvale.***

At Webster, for ***Hall's Valley, St. Johns, Peru, Montezuma,*** and all Mining Points on the Snake River.

At Como, with Stages for ***Hamilton, Breckenridge, Lincoln City, Conger, Georgia Gulch*** and all Mining Points on the Blue, Grand, and Eagle Rivers.

At Red Hill, for ***Fairplay, Alma, Dudley, Montgomery, Mosquito*** and ***Leadville.***

At Buena Vista, with Daily Stages for ***Granite, Twin Lakes, Malta, Leadville, Ten Mile, Red Cliff and Roaring Forks.*** Also for ***Alpine, Pitkin, Crested Butte, Gunnison City, Ouray, Lake City,*** and all the great *MINING CAMPS* of the **Gunnison Valley** and ***Dolores.***

NOTICE.—The **DENVER, SOUTH PARK and Pacific R. R.** is now completed to GRANITE, and will be running its Trains to

MALTA AND LEADVILLE,

ABOUT JULY 5, 1880.

DAVID S. DIGERNESS COLLECTION

AN EIGHT-CAR SPECIAL passenger train of the Denver, Leadville & Gunnison was in Morrison one summer day in 1895. The train was being hauled by locomotive

Number 214 — originally DSP&P Number 65 — a 2-8-0 built in 1883 by the Cooke works. Lettering on the engine and cars indicated ownership of the Old South Park by the Union Pacific at this time, despite the DL&G name. On the far side of the train the roofs of two boxcars can be seen, while the old turntable is in plain sight.

W. G. CHAMBERLAIN PHOTO
RICHARD RONZIO COLLECTION

THE MORRISON BRANCH was new in the late 1870's, when this photograph was taken by the renowned William G. Chamberlain. DSP&P locomotive Number 1, the "Fairplay," was pulling a mixed train out of town. This locomotive was built in 1873 by Dawson & Baily, and had a 2-6-0 wheel arrangement.

Gulch — as well as the distant Gunnison country and the Blue River tributaries in Summit County. The latter two areas are located on the west side of the Great Divide.

Teamsters doing the freighting to Webster from all these locations had more business than they could handle. The railroad, in turn, was picking up the business from the freight lines, as well as all business from a number of stage lines, providing less expensive transportation services to and from Denver.

Jesse noted that the mouth of Halls Valley was reached by the railroad early in January of 1879, where a frame depot and water tank had been constructed, and a town had formed at the end-of-track — called Webster — at an elevation of 9,120 feet. The great activity of roadbed construction brought on by Leadville's silver rush occurred just beyond Webster during the winter of 1878.

Jesse was informed that the railroad construction force was at Webster when the Leadville mining boom caused the South Park line to put tremendous effort into pushing the railroad up the hill. Also, it was at this point in the railroad's existence that DSP&P officials decided that Leadville was to be one of their main points of destination, their goal being to tap that great and fast-growing bonanza mining camp. Two new railroads were organized to run from Webster to try to achieve this aim one way or another.

With the idea in mind of finding the best and quickest route from Webster to Leadville, the Denver, South Park & Ten Mile Railroad was incorporated by Evans and his associates on December 13, 1878. The projected road was to run, by the most feasible route, from a connection with the DSP&P in Park County into Summit County on the Blue River; then the line would follow the river down to its confluence with Ten Mile Creek, thence up along the Ten Mile to its headwaters. From there, the route would go into Lake County, at the head of the valley of the Arkansas River, and on up into Leadville.

Notice Excursion Rates and Study this Map.

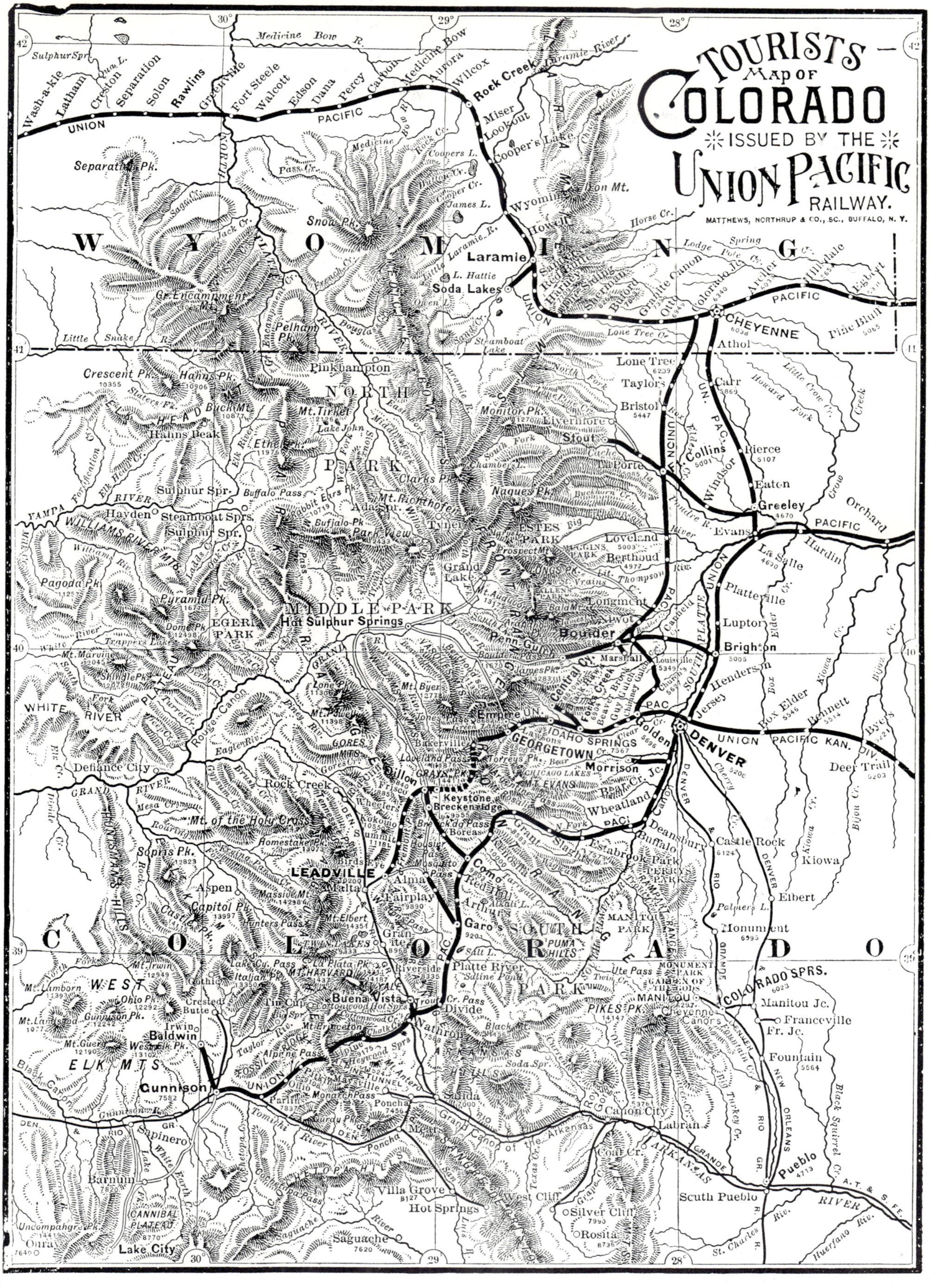

Leadville and Aspen are reached by the South Park Division.

For Tickets and Rates apply to B. P. M. Kimball, 1703 Larimer st., or at Union Depot Ticket Office.

The Union Pacific is the Greatest Tourist Line. This Map Proves it.

DAVID S. DIGERNESS COLLECTION

THIS WAS MORRISON during the twilight years of operations on the South Park line. The year was 1922, four years after regular passenger runs had been discontinued. However, the C&S ran tri-weekly mixed service as late as 1924. Gasoline pumps had arrived on the sidewalks in front of Morrison's stores; nevertheless, the

main street still was relatively free of automobile traffic. Unfortunately, the railway facilities were falling into disrepair. Ruins of the turntable pit can be seen below the depot, adjacent to the gondolas. It might also be noted that Red Rocks Park was yet to be developed.

DAVID S. DIGERNESS COLLECTION

THE OFFICE of the quarry operations of the Morrison Stone, Lime & Townsite Company was formed on October 24, 1872, by organizers of the Denver, South Park & Pacific Railway, along with George M. Morrison.

STACKS OF LUMBER, stone and limestone waited at the end-of-track on the Morrison Branch in 1874. Bear Creek winds its way through the center of Morrison, where a bridge had been built for the convenience of persons journeying to and from the South Park area.

W. G. CHAMBERLAIN PHOTO
DAVID S. DIGERNESS COLLECTION

W. G. CHAMBERLAIN PHOTO
RICHARD RONZIO COLLECTION

TURKEY CREEK winds its way out of the Foothills of the Front Range a few miles south of Morrison. The wagon road in this view was built beside the creek to permit access to Foothills towns and also was a gateway to South Park. This photograph probably was taken during the late 1870's.

F. E. DuBOIS PHOTO
DAVID S. DIGERNESS COLLECTION

COLORADO CAMERA CLUB members had a delightful time while on the 1892 railway excursion to Morrison, which included a side-trip to the Garfield Quarry. Their motive power was DL&G Mason bogie 57, named "Buena Vista," a trim little 2-6-6T built in 1880.

LOCOMOTIVE NUMBER 108 of the Union Pacific, Denver & Gulf Railroad, derailed in the late 1890's while backing downgrade on the Garfield Quarry Branch. The engine, originally Colorado Central's Number 15, was built by the Cooke works in 1884. She later became Colorado & Southern Number 13. The photographer was looking north, near the switch at Bear Creek, adjacent to the Hogback.

RICHARD RONZIO COLLECTION

DAVID S. DIGERNESS COLLECTION

ANOTHER MISHAP involving the 108 took place on the Morrison line during the 1890's. While the UPD&G locomotive was backing a freight train upgrade past the Morrison depot — near the end-of-track — two boxcars derailed, almost overturning into Bear Creek.

The other railroad, incorporated by Evans on the same date, was the Denver, South Park & Leadville Railroad. The projected line of this road was to be by the most feasible route, from a point of connection with the original survey of the Denver, South Park & Pacific Railroad, further along on Trout Creek, then up along the valley of the Arkansas River to Leadville.

The original location survey line was run to Trout Creek, after crossing South Park, and surveyed beyond to the Pacific side of the Sawatch Range, with the intention of serving the Gunnison and San Juan regions, and the line was to run from there west into Utah, connecting with the Central Pacific. With this idea definitely in mind, the original location survey was chosen and adopted. This line of reasoning was a deciding factor in adopting the route of the Denver, South Park & Leadville Railroad. However, this line was to be a branch of the Denver, South Park & Pacific Railroad, which would follow the original location survey, connecting at the mouth of Trout Creek; thence the branch would run upgrade along the Valley of the Arkansas River to Malta, then east four miles to Leadville. So, the two newly incorporated "paper railroads" were short-lived, having been abandoned in favor of a branch-line connection of the DSP&P to tap the Leadville area.

Jesse — as will be brought out — was a collector of minerals, and, while in Webster, he met a man by the name of General Craig. Craig invited him to visit the Radical Mine to see the operation and get specimens of galena and gray copper, which were assaying an average 60 ounces of silver and 60 percent lead. Craig was general manager, and the mine was owned by the Montezuma Silver Mining Company. They traveled by wagon road over Handcart Pass into Summit County.

Just the year before, the company built the wagon road on an easy grade to Webster on the DSP&P, 12 miles distant. Jesse got his specimens right out of the mine on a tour, and he visited the company's new concentrating mill during the same tour. He spent the night as a guest of Craig, returning to Webster the following morning.

On to the Carbonate Camp

From Webster, Jesse took the stagecoach on to Leadville, stopping enroute at Fairplay overnight at a cost of $3.00 (including three meals and lodging). The next morning, after a

OTTO C. PERRY PHOTO
DAVID S. DIGERNESS COLLECTION

AT THE END OF THE SILICA BRANCH a mineral train was being switched by a C&S double-header during 1941. This branch technically was abandoned at the time Otto Perry took this photograph; however, mining activities in the area provided enough carloads to make up a train now and then — such as this one about to depart for Denver, with engines 69 and 70 at the head-end. The C&S continued to operate this branch-line until late 1942. **THE WATERTON FILTER BEDS** of the Denver water system appear in the background of this late-afternoon photograph, taken on March 31, 1941. The entrance to Platte Canyon is just around the curve in the distance (to the right). This C&S train was Denver-bound with a string of 28 freight cars.

OTTO C. PERRY PHOTO
DAVID S. DIGERNESS COLLECTION

JOHN W. MAXWELL PHOTO

THIS CLOSE-UP VIEW of Colorado & Southern locomotive 70 was taken on May 17, 1941, near Chatfield. While heading southbound toward Waterton, this train had been delayed by the derailment of a truck wheelset.

good breakfast, he continued on to Leadville. After a fast 30-mile stage trip from Webster the day before, Leadville was still 40 miles away by way of Weston Pass. The going was slow because of a loaded wagon train on the pass, which slowed them down to a snail's pace. But the lost time was made up on the remainder of the trip. Upon arriving in Leadville, he located a room in the new Grand Hotel and immediately took a tour of the area. Besides observing Oro City, the town for which the DSP&P named their Number 3 locomotive (which he saw in the enginehouse at Platte Canyon), he got a closeup look at California Gulch, which he found to be the liveliest area that he had ever seen. The town of Leadville was even more a marvel of activity. The entire scene showed off the area's great mining boom. The appearance was a duplication of what the Georgetown Mining District had earlier been — but to a greater degree.

Jesse got around pretty well it seems, since he ran across some friends who had earlier worked in the Georgetown area mines. This got him going in the right direction, for they, in turn, referred him to individuals with whom he was able to talk at great length, which allowed him to obtain the inside information he wanted. He was told of all types of great wealth. It so happened that Jesse picked just the right time to take his first-hand observations of Leadville, for this was when he was introduced to H. A. W. Tabor, who just happened to come along. At the time, "Haw" Tabor was the Lieutenant Governor of Colorado, and just the year before (in 1878), had been Leadville's first mayor. Horace Tabor — the original "Silver King" of Leadville — took time from his business affairs to provide Jesse with a considerable amount of information about the town and himself, and they became well-acquainted. From this time on, just about every time Jesse was in Leadville, Tabor happened to be there and took time to talk with him. They even got together in Denver a number of times, as well as at Georgetown.

Jesse's records about H. A. W. Tabor — Leadville's most famous character — and of the entire area are most dramatic.

Tabor and his wife, Augusta, had settled on a farm in Kansas, after having been married in the state of Maine. In 1859, Horace, his wife and their new son took their cattle and a team of oxen and journeyed by covered wagon to the Pikes Peak Region. Augusta was the eleventh woman to arrive in the frontier settlement of Denver City. After camping there for a week, the Tabors moved to the west and camped along Clear Creek at the foot of the mountains, where Golden now stands. Tabor left his wife in camp, with six months of provisions, to take care of their son and the cattle while he entered the mountains with a couple of men to look for gold. However, he was unsuccessful and returned to camp. Then, packing up his belongings, he and his family moved farther into the

OTTO C. PERRY PHOTO
DAVID S. DIGERNESS COLLECTION

COLORADO & SOUTHERN locomotive Number 70 was making up a train load of minerals at Silica, on the branch out of Waterton, in this view of March 31, 1941. An old-style harp switch stand, still in use, can be seen beside the train.

OLD SOUTH PARK MAP, DENVER TO WEBSTER

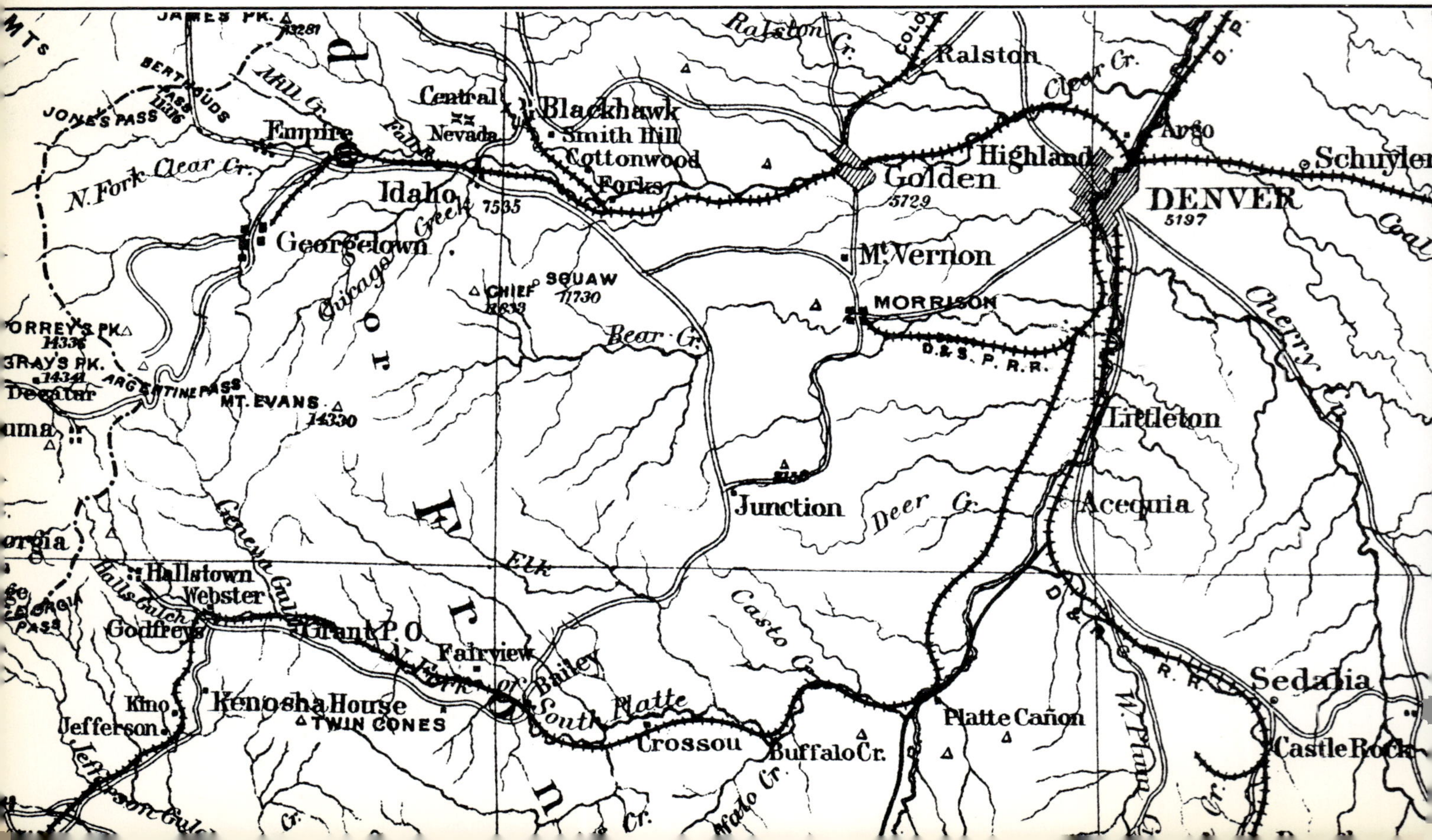

mountains to Payne's Bar on Clear Creek, where he made his first mining claim. Other miners advised Tabor to move his family out of the mountains for the winter because of the danger of snowslides. Tabor moved his wife and son back to Denver. He then returned to find his claim had been jumped, and that he had been tricked into leaving by the snowslide story. No laws existed to protect his interest at that time.

After returning to Denver, Tabor's wife cooked for boarders throughout the winter and took care of their son. Then, traveling to Colorado City, Tabor built a business house east of there and tried to promote the location as an ideal site for building a town. Returning to Denver, he tried his best to get people to go there and establish a town, but without success. Following this failure, Tabor moved his family to the Colorado City area — now part of Colorado Springs — which consisted of only a few log houses at that time. The town company gave them some lots in honor of Tabor's wife being the first woman to arrive there.

In search for gold, Tabor moved his family into the mountains over Ute Pass and across South Park to Cache Creek, where the town of Granite later was established along the Arkansas River. He whip-sawed some lumber to make sluice boxes, and put them in place along the creek. Gold! There was plenty of it to be had! However, not knowing how to separate the gold from the sand without quicksilver, he had to abandon the site.

Tabor — with his family and two other men, Nathaniel Maxey and S. P. Kellogg — was the first to live at the Granite site. Having camped there one month with little success, they moved on up the Arkansas River. Granite later turned out to be very rich property.

It was reported that gold had been found in California Gulch, but supplies were needed to extract it. This was

W. G. CHAMBERLAIN PHOTO
STATE HISTORICAL SOCIETY OF COLORADO

NEW DSP&P IRON RAIL was laid on a roadbed blasted out of solid rock in 1878. In portions of the Platte Canyon, it was necessary to support the line on retaining walls rather than blasting through the granite cliffs.

W. H. JACKSON PHOTO
DAVID S. DIGERNESS COLLECTION

THE GRADE OF THE DSP&P once entered the Platte Canyon at this point. This photo, taken late in the 1890's, shows the High Line Canal beside the Platte River and the DSP&P trackage to the left. The canal was built for irrigation purposes and to supply water for the City Park lakes in Denver.

JOHN W. MAXWELL PHOTO

ON NOVEMBER 15, 1941, workmen were busy dismantling the C&S bridge over the Highline Canal — on the branch which led to Silica. The entrance to Platte Canyon is in the background. It looks as though a "brass hat" was watching his workmen on this unusually warm autumn day.

Tabor's chance. They went up to the gulch and found Slater and Abe Lee panning an average of a dollar to the pan. Tabor then slaughtered the cattle they had driven in from Kansas and divided the beef among them. In the meantime, other supplies had been sent for and were brought in by Mexicans, who, in turn, helped to build a cabin in which Tabor's family could live that summer. The Tabors arrived in California Gulch on May 8, 1860, and by 1861 — when the diggings were exhausted — they had acquired what they considered quite a little fortune — about $7,000. From there they went over the Mosquito Range to the town of Buckskin Joe, named for a man who discovered the first mine. His name happened to be Joe and he wore buckskin clothes!

There was great excitement, and mines boomed with riches; hundreds of people were there and Tabor had gone over with the rush. There he started a store and kept the post office, while his wife kept a boarding house. These proved to be profitable ventures for the Tabors, as most people arrived in camp with nothing but packs upon their backs. The Tabors had a house and enjoyed a period of financial stability and security. After six years of being postmaster — and even mayor for awhile — the mines all seemed to play out. Tabor and his family returned to California Gulch and opened a place in which to continue in the mercantile trade. He also became the new postmaster there and ran an express office as well.

At this time, one William Van Brooklyn packed mail and express into the California Gulch district. He was paid 75 cents for each letter he packed in, and was paid accordingly for any express shipments he could bring in on his mules.

That summer, the Tabors weighed most of the gold that was taken out of the upper end of the gulch. There was many a miner who knew nothing about weighing gold.

A man named Green took the contract to supply the first 300,000 ties to the AT&SF Railroad. All of these ties were cut in California Gulch. He owed Tabor a good deal of money for supplies and Tabor became alarmed when he found out that the man was not going to make a success of it. So, he took the contract off Green's hands. He had hoped to get his money back, but lost money doing it. It took Tabor five months to transport those ties down the mountain to the Arkansas River. Tabor expected to float them down when the water was high

MORRISON SMITH PHOTO
JOHN W. MAXWELL COLLECTION

C&S LOCOMOTIVE Number 69 and a combination baggage-chaircar were in service on November 15, 1941, dismantling the bridge over the Highline Canal, located on the Silica Branch. In this view, the photographer was looking northwest toward Platte Canyon. To the right is the site of today's Martin-Marietta rocket plant (behind the silo).

in the spring, but there was very little snow that season, which kept the runoff low and forced him to haul the ties by teams to the river. After Tabor got through with the tie business, he realized that he had worked two years and had not made a dollar. During the time Tabor ran the tie business, he worked right along with the men he employed, while his wife ran the store for him, providing enough money to enable him to get the ties through.

Late in June of 1877, a grocer by the name of Charles Mater came up from Granite, which, at the time, was the county seat of Lake County. Mater constructed the first building in the original townsite of Leadville, and opened for business with a stock of groceries. Several other log cabins sprang up even before Mater supplied his store. In the meantime, excellent mines began to develop in the nearby hills, and settlers slowly drifted in from the older mining camps.

After coming back from the town of Buckskin Joe, Horace Tabor re-established a store in the upper end of California Gulch, at the new Oro City, and operated it there for many years. However, during the summer of 1877, Tabor moved his stock of goods to Leadville and set up one of Leadville's first mercantile stores.

..., see Union Pacific, Denver & Gulf Ry.

DENVER TO ALMA, BUENA VISTA, GUNNISON AND BALDWIN.

31	21	23	Mls	*June* 13, 1897.	Elev.	32	22	24
A. M.	A. M.	P. M.		LEAVE] [ARRIVE		P. M.	P. M.	A. M.
*8 15	*8 15	†6 30	0	..........**Denver**..........	5170	6 05	6 05	4 00
d1 20	*1 35	†2 15	88	lve.........**Como** ♁.... arr.	9775	1 05	12 40	7 15
1 35	1 55	2 37	94	..........Red Hill	9517	12 50	12 20	6 50
	2 30	3 08	103	..........Arthur's..........	9162	12 30	11 45	6 10
2 00	2 40	3 20	105	arr..........**Garos**......lve.	9172	12 25	11 35	6 00
	3 30		115	..........Fairplay........♁	9886		10 ·7	
	4 00		121	arr......+**Alma** ♁......lve.	10228		*10 20	
2 00	P. M.	3 20	105	lve.........**Garos**.......arr.	9172	12 25	A. M.	6 00
2 22		3 55	114	.........Platte River.........	8934	12 03		5 20
2 45	25	4 30	120	..........Bath..........	9460	11 45	26	4 53
3 05	P. M.	5 05	127	..........McGees..........	8648	11 24	A. M.	4 10
3 23	†2 30	5 40	133	arr.. **Schwander's**....lve.	7814	11 04	6 30	3 26
3 38	2 50	6 00	137	arr..+**Buena Vista** ♁...lve.	7943	v10 52	†6 10	†2 55
d3 38	†2 55	†6 10	137	lve....**Buena Vista**....arr.	7943	10 52	6 00	2 50
3 50	3 15	6 50	133	lve....**Schwander's**....arr.	7814	10 40	†5 40	2 30
4 14	P. M.	7 40	142	......Mount Princeton......	8169	10 16	A. M.	1 40
4 40		8 25	149	..........Alpine..........	9247	9 51		1 00
5 01		9 00	153	+..........St. Elmo........♁	10041	9 34		12 30
5 12		9 25	156	..........Romley..........	11005	9 25		12 15
5 38		10 25	162	.........Alpine Tunnel......	11596	9 00		11 25
6 20		12 01	175	+..........Pitkin..........♁	9180	8 08		9 40
6 38	31	12 30	181	..........Ohio City..........	....	7 50	32	9 00
6 59	P. M.	1 03	190	..........Parlin's..........	7911	7 30	P. M.	8 20
d7 30	c2 45	1 45	202	+........**Gunnison**.......♁	7649	v7 00	6 00	†7 30
P. M.	4 05	P. M.	217	.........Castleton..........	8377	A. M.	4 50	A. M.
	4 20		219	arr.........**Baldwin**......lve.	8686		c4 30	

DENVER TO MORRISON.

Sun.	9	15	7	M	*Jun.* 13, 1897.	Elv.	8	16	10
A. M.	P. M.	A. M.	A. M.		LEAVE] [ARRIVE		A. M.	P. M.	P. M.
§8 30	†4 10	†11 0k	†8 00	0	..**Denver** ♁	5170	10 00	3 20	6 15
.. ..	.. ..	11 10	.. ..	1	..West Denver..	5187	.. ..	3 15	
8 45	4 23	11 20	8 15	3	..Valverde..	5214	9 48	3 02	6 03
8 56	...	11 32	..	7	..Mooreville...	5282	...	2 50	...
8 58	4 35	11 40	8 28	8	.**Sheridan Jn.**	5284	9 35	2 45	5 50
9 01	4 40	11 45	8 31	9	..**Fort Logan**..	5311	9 30	2 38	5 45
9 06	4 45	11 55	8 36	11	...Bancom...	....	9 25	2 30	5 40
9 25	5 05	12 25	8 55	17	..**Morrison.** ♁	5753	†9 05	†2 00	*5 20
A. M.	P. M.	NO'N	A. M.		ARRIVE] [LEAVE		A. M.	P. M.	P. M.

* Daily; † daily, except Sunday; *a* Monday only; *b* daily, except Saturday and Sunday; *c* Monday, Wednesday and Friday; *d* Wednesday and Saturday; *k* Saturday only; *v* Monday and Thursday. + Coupon stations; ♁ Telegraph stations. STANDARD—*Mountain time.*

For Map, see Page 572.

W. G. CHAMBERLAIN PHOTO
STATE HISTORICAL SOCIETY OF COLORADO

DEANE'S TIE CAMP AND SAWMILL were located at Milepost 26.8 in the Platte Canyon. Nearly 30,000 ties, each approximately seven by seven inches and seven feet long, were floated down river for use on the railroad grade. More than 100,000 ties came from this tie camp for the DSP&P railroad. Deane's multi-story brick hotel is under construction in this scene.

STATE HISTORICAL SOCIETY OF COLORADO

SHORT STRETCHES OF TANGENT TRACK were found on the DSP&P in the Platte Canyon just above the first iron truss bridge. This view looks up river to the west, showing a construction tent beside the newly laid rail in 1878.

W. H. JACKSON PHOTO
DAVID S. DIGERNESS COLLECTION

DEANSBURY STATION in Platte Canyon, as photographed by W. H. Jackson in 1884, shows the newly completed hotel built by Mr. C. A. Deane. His tie-camp and sawmill both were located at this point, providing construction materials for the DSP&P. Deane's logging camp was up the gulch to the south. Later, this station became known as Strontia Springs.

DEAN'S STATION — also called Deansbury — was ready for business in 1884, with fresh paint on the new porch balcony. A Mason bogie, equipped with a Congdon smokestack, was on a siding, enroute to Denver. In later years, this brick hotel became a summer resort, one of the many small resorts along the South Platte River.

CHARLES WEITFLE PHOTO
DAVID S. DIGERNESS COLLECTION

JOHN W. MAXWELL PHOTO

DURING THE AFTERNOON of May 17, 1941, C&S Number 70 drifted downgrade past Strontia Springs (Deansbury) with only a caboose. During mid-summer, the lower reaches of Platte Canyon can be quite warm — compared with the high-mountain passes of the Centennial State.

Denver, South Park & Pacific

RAILROAD.

TOURISTS.

The Day Express, leaving Denver at 7.30 A. M., affords an excellent opportunity for Tourists to visit the wonderful

PLATTE AND BLACK CANONS,

Dome and Cathedral Rocks and

KENOSHA SUMMIT,

10,139 Feet above the Sea.

There are also several hours in which to enjoy the

FINE TROUT FISHING

along the line of this road before the return train, which reaches Denver at 8.05 P. M. See time card on preceding page.

EXCURSIONS.

This line is acknowledged to have the finest Excursion points of any in Colorado. The many thousands who have visited its

WILD AND BEAUTIFUL SCENES,

will bear us out in the assertion that nowhere are there finer resorts for Pic-nics and Moonlight Excursion Parties.

Special Rates will be made and Special Trains Run for Parties,

On application to the General Ticket Agent.

PHOTOS BY JOHN W. MAXWELL

INTREPID RIDERS and photographers of the iron horse, John Maxwell and Dick Kindig had the honor of being aboard one of the last runs of the C&S up the South Platte River on May 17, 1941. Nine empty cars were taken to South Platte station and dropped before noon for flourspar loading. Engine 70 then was wyed for the return trip. In this scene — looking over the cab and tender of the locomotive — it can be seen that a back-up light had been outfitted on the 2-8-0.

A NEW PAINT JOB graced this locomotive, shortly after being purchased from the Kansas Central Railway. During the summer of 1877 this engine was kept busy hauling construction trains in Platte Canyon, as this scene depicts. Built by Dawson & Bailey in 1874, this 4-4-0 was named the "Platte Canyon" by the DSP&P.

HAWKINS PHOTO
DAVID S. DIGERNESS COLLECTION

W. H. JACKSON PHOTO
RICHARD RONZIO COLLECTION

THIS PRINT was made contact size from one of W. H. Jackson's original glass-plate negatives, which he produced while on a trip through Platte Canyon in 1887. Locomotive 42 was one of the 19 engines of this type built by the Mason Locomotive & Machine Company of Taunton, Massachusetts.

OPPOSITE: DSP&P Number 10—the "Granite"—was beautifully recorded on a glass-plate negative by W. H. Jackson in 1879. The engine was on the Deansbury Bridge—simmering softly to herself—as the foremost of the American West's photographers "took her picture" with a twin-lens stereographic camera. When viewed through a stereoscope (or "Stereopticon"), the scene appeared to be three-dimensional, and Jackson sold thousands of early Colorado railroad views of this type before the turn of the century. The "Granite" was almost new when the great master photographer captured this view on glass—the 2-6-6T having been delivered from the Mason works the year before.

W. H. JACKSON PHOTO
DAVID S. DIGERNESS COLLECTION

DAVID S. DIGERNESS PHOTO

OPPOSITE: Part of Denver's drinking water rushes toward the photographer in this view looking up-river on the South Platte River. The Old South Park bridge still stands at this location, largely used by crews of the Denver Water Department and sports fishermen. This bridge is just west of Deansbury (later named Strontia Springs) near Milepost 27. This bridge replaced the original Deansbury Bridge.

THIS HUGE BOULDER, resting in the waters of the South Platte River, served as a landmark to locomotive engineers of the Old South Park. The long-abandoned track grade can be seen on the right. This color photograph was taken from nearly the same location as the following color view. In this scene, the author was looking due west, while the following view was recorded as he faced the opposite direction.

DAVID S. DIGERNESS PHOTO

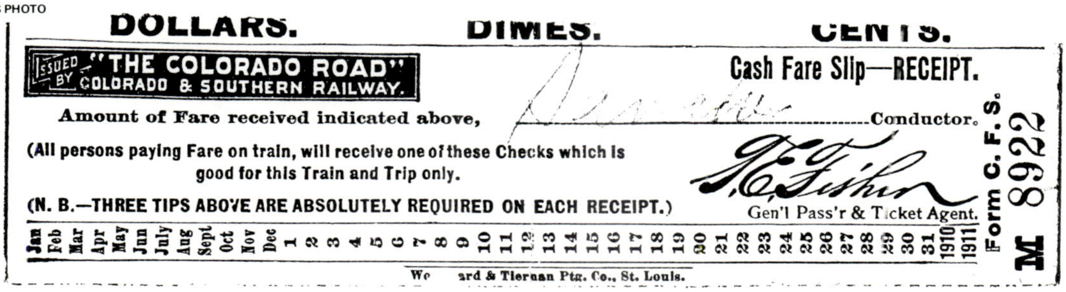
DOLLARS. DIMES. CENTS.

ISSUED BY "THE COLORADO ROAD" COLORADO & SOUTHERN RAILWAY.

Cash Fare Slip—RECEIPT.

Amount of Fare received indicated above, ...Conductor.

(All persons paying Fare on train, will receive one of these Checks which is good for this Train and Trip only.

(N. B.—THREE TIPS ABOVE ARE ABSOLUTELY REQUIRED ON EACH RECEIPT.)

Gen'l Pass'r & Ticket Agent.

Form C. F. S.

M 8922

Jan Feb Mar Apr May Jun July Aug Sept Oct Nov Dec 1 2 3 4 5 6 7 8 9 10 11 12 13 14 15 16 17 18 19 20 21 22 23 24 25 26 27 28 29 30 31 1910 1911

Wo[illegible]ard & Tiernan Ptg. Co., St. Louis.

GRIFFITH & GRIFFITH STEREOPTICON CO.
DAVID S. DIGERNESS COLLECTION

CHARLES W. WEITFLE STEREO CARD
DAVID S. DIGERNESS COLLECTION

OPPOSITE: This classic sepia-tone photograph was produced for the Griffith & Griffith Stereopticon Company during the late 1880's. It features a DSP&P freight train heading downgrade in Platte Canyon. The wealth of goods on this train consisted of livestock, lumber and sacked ore concentrate for the markets of Denver and the East. Hauled by Number 112—a 58,300-pound 2-6-0 built by the Cooke Locomotive & Machine Company in 1884—the train was crossing the famous Deansbury Bridge on the narrow-gauge mainline of the Old South Park.

SOON AFTER THE CONSTRUCTION of the DSP&P's Platte Canyon line in 1879, pioneer photographer Charles Weitfle recorded this mixed train heading west in the canyon. It was powered by Mason bogie Number 8—the "Lake City"—which later was given the number 44. This 2-6-6T tank-type locomotive was scrapped in 1889, after only a mere 10 years of service.

DAVID S. DIGERNESS PHOTO

OPPOSITE: The DSP&P grade followed the north bank of the South Platte River, as viewed in this color photograph. This point is a few miles below the site of South Platte station. A huge boulder (shown on page 151) nearly blocks the flow of water in this part of the gorge—also shown in a W. H. Jackson view taken in 1886.

THE LONG-ABANDONED MAINLINE of the DSP&P ran in front of the South Platte Hotel, shown in the recent color view above, portraying the old townsite of South Platte. This station also was the junction for the Night Hawk Branch, which headed south at this point. The South Platte River flows just behind these buildings, entering the canyon to the left. From South Platte, the mainline paralleled the North Fork of the South Platte westward to Pine, Bailey and Grant; thence the route traversed Kenosha Pass to reach South Park and the land beyond—the highest part of Colorado.

DAVID S. DIGERNESS PHOTO

RICHARD B. JACKSON PHOTO
DAVID S. DIGERNESS COLLECTION

O. WESTERMAN PHOTO
DAVID S. DIGERNESS COLLECTION

OPPOSITE, ABOVE and BELOW: The last passenger train in Platte Canyon, April 10, 1937, was photographed west of Strontia Springs, beside the South Platte River. Eastbound locomotive Number 71 was pulling a trainload of passengers who were saying farewell to Colorado & Southern service through this rugged mountain gorge. The depot in the scene below was at South Platte station, departure point for the Night Hawk Branch. Posing here in the 1890's, two women were waiting on the platform for the afternoon passenger train. The platform was cluttered with boxes and wooden kegs awaiting shipment. The mainline enters Platte Canyon to the east of the depot, and a sign points the way to Deckers by way of the bridge (to the right).

BY THE 1950's, these broken pilings were all that remained of a bridge on the Night Hawk Branch of the Colorado & Southern Railway. This bridge crossed the South Platte just below the town of South Platte. The four-mile line, built to serve the Cheesman Dam construction project and the feldspar mines around Night Hawk, was abandoned in 1916.

WALLS OF GRANITE lined each side of the Platte Canyon and dwarfed the steel trail of the Denver, South Park & Pacific Railroad. An early survey had falsely declared that this route was "impenetrable." Waycar 72, pulled by Number 42, carried the great W. H. Jackson across the new railway in 1886, enabling him to record the wondrous beauty of this line on glass negatives. These views were used by the railway company to publicize tourist travel—and Jackson also offered sepia-tone prints and stereopticon view cards through commercial outlets. Incidentally, the term "stereopticon" was the one generally used in Colorado to mean a double stereoscope view. Nearly every household in the state had a stereoscope viewer during the period Jackson became famous.

DAVID S. DIGERNESS COLLECTION

W. H. JACKSON PHOTO
DAVID S. DIGERNESS COLLECTION

THE ROCK FORMATION known as Rolling Stone was a well-advertised landmark on the Denver, South Park & Pacific — situated beside the tumbling South Platte River. Except for a change in vegetation, this scene has changed very little since the days of railway traffic through Platte Canyon.

OVERLEAF: During the return part of his 1886 expedition over the South Park line, W. H. Jackson instinctively realized the fine pictorial qualities—in summer afternoon light—of the gnarled sidewalls and the huge, rounded granite boulders typical of the lower reaches of South Platte Canyon. Using waycar 72 as his rolling darkroom, he stopped his special train at Rolling Stone Curve long enough to record this outstanding view.

W. H. JACKSON PHOTO
STATE HISTORICAL SOCIETY OF COLORADO

DAVID S. DIGERNESS PHOTO

42

W. H. JACKSON PHOTO
DAVID S. DIGERNESS COLLECTION

OPPOSITE: Dome Rock—one of many famous landmarks on the DSP&P—was a favorite picnic ground for train-loads of passengers. When trains were still operating, a footbridge provided access to the far side of the river.

DOME ROCK STATION featured a small, simply designed frame depot at the foot of Dome Rock. This served as a combination depot and living-quarters. The massive stone formation towering above the station — along the North Fork of the South Platte — was one of the "views of grandeur" on the South Park line mentioned in Jesse Randall's journal. (See the text.)

DAVID S. DIGERNESS PHOTO

GEORGE E. MELLEN PHOTO
DAVID S. DIGERNESS COLLECTION

DOME ROCK is a landmark formation in Platte Canyon. This solid-granite formation is characteristic of rock formations found throughout the Tarryall Mountains and along the South Platte River. The Old South Park utilized its value in their tourist advertising — along with mention of the numerous other scenic attractions in the Platte Canyon area. Pioneer photographer George E. Mellen caught DSP&P engine 112 in front of Dome Rock in the late 1880's. This 2-6-0 was built by Cooke in 1884 and was hauling a livestock train into Denver.

CAMP AND PLANT CF&I COLLECTION

DOME ROCK DEPOT

DAVID S. DIGERNESS COLLECTION

FAMOUS DOME ROCK

DAVID S. DIGERNESS COLLECTION

CHEESMAN DAM, on the South Fork of the Platte River, was nearing completion when this photograph was taken, October 8, 1903. Notice the size of the workers on top of the wall, compared with the quarried stone used in the construction.

CHARLES WEITFLE PHOTO
STATE HISTORICAL SOCIETY OF COLORADO

OPPOSITE: Denver's Cheesman Dam was under construction when this photograph was taken in 1903. This view shows the site of the stone quarry used during the building of this dam-and-reservoir project. The quarry was within the boundaries of the reservoir site and this area now is under water. "Polly Pry"—Mrs. Lionel Ross Anthony, one of Colorado's celebrated women journalists and a prominent socialite—was captured on film as she took a ride in a sling suspended over the edge of the canyon.

STATE HISTORICAL SOCIETY OF COLORADO

"ROLLING STONE CURVE" in Platte Canyon attracted a great deal of attention during the early days of the DSP&P. A Mason bogie locomotive was heading upgrade to Webster with a trainload of supplies and passengers in this view of 1878.

W. H. JACKSON PHOTO
DAVID S. DIGERNESS COLLECTION

STATE HISTORICAL SOCIETY OF COLORADO

ABOVE: Posing for the photographer, the crew of this W. H. Jackson special of 1886 held their stance long enough to record this classic view. The blurred foliage in the foreground is evidence of the slow-acting emulsion used on the early glass-plate negatives. The little waycar is an earlier version of one of the Denver, South Park & Pacific 4-wheel cabooses. Notice the heavy sisal rope draped over the roof of the waycar for some unknown reason.

A FIVE-CAR WESTBOUND FREIGHT TRAIN had pulled to a halt below Cathedral Rock on the mainline of the DSP&P. The Old South Park was short of cabooses during the boom years of construction, which sometimes overfilled the locomotive cab with train crew.

STATE HISTORICAL SOCIETY OF COLORADO

OPPOSITE: The Cathedral Rocks in Platte Canyon formed the backdrop for this special W. H. Jackson train. Mason bogie 42, a newly polished 2-6-6T tank-type engine, was coupled to a new DSP&P coach, Number 9, sporting its original coat of varnish. This was during the summer of 1887 along the mainline of the South Park line. A controversy exists among photographers and historians as to whether W. H. Jackson or George E. Mellen photographed the views of this trip over the line, since Mellen's credit line appears on prints also attributed to Jackson. However, the original 20x24-inch glass negative of this print—now in the possesion of the Colorado State Historical Society has additional space at the bottom of the view. In that space, only Jackson's print number and name are enscribed. A reproduction from this negative clearly showing this inscription appears in Sundance Publications' limited edition, William Henry Jackson's "Rocky Mountain Railroad Album." It may have been that Mellen purchased the photographs in question from Jackson and sold prints under his own name, but the actual facts may never be known.

42

W. H. JACKSON PHOTO
DAVID S. DIGERNESS COLLECTION

OPPOSITE: This W. H. Jackson view is a reproduction from the original negative made by the great photographer on a summer day back in 1887. This illustrates the exquisite detail obtainable from Jackson's large-format glass-plate negatives recorded almost 100 years ago.

W. H. JACKSON PHOTO
DAVID S. DIGERNESS COLLECTION

TWO MASON BOGIES MET one summer day in 1879 at the town of Buffalo. A mixed passenger train heading west was holding the main line of the DSP&P, while a work train waited clearance to proceed on into Denver. The rock formation in the background was called Tower Mountain by the local people. This photograph was taken from the Buffalo depot.

CHARLES WEITFLE PHOTO
DAVID S. DIGERNESS COLLECTION

A MASON BOGIE 2-6-6T locomotive, with a mixed freight-and-passenger train, was pictured in Platte Canyon by pioneer photographer Charles Weitfle. At this point, the gorge widens and a good view could be obtained of Number 6, the "Ten-Mile." This engine was a tank type, i.e., the bunkers for fuel and water were an integral part of the engine, rather than being contained in a separate car connected to the engine. She is shown here in the early 1880's.

DAVID S. DIGERNESS COLLECTION

A MASON BOGIE named "Ouray" — a 2-6-6T equipped with a diamond stack — and freight locomotive Number 191 — with its Congdon stack — are pictured at Pine Grove station in October of 1885. Locomotive 46 was built by Mason in 1879, and the 191, on the wye, was built by Baldwin in 1880. The 2-8-0 now is on display at the famous Colorado Railroad Museum in Golden.

IN SEPTEMBER OF 1901, heavy rains softened the C&S roadbed west of Foxton and Mogul Number 10 took a spill. The train was eastbound on an excursion run to the popular resort of Crystal Lake.

TOM ST. JOHN PHOTO
DAVID S. DIGERNESS COLLECTION

ENGINEER Whitney, Fireman Burke, Conductor Smith and the head-end brakeman took a rest for the photographer while stopped at Buffalo in the mid-1880's. This DSP&P narrow-gauge locomotive, Number 112, was built by Cooke in 1884.

ENGINEER PATRICK "CURLY" COLLIGAN poses with his conductor around 1910. Curly was — literally and figuratively — a big man on the South Park. He began working for the road in 1887 and ran the last passenger train to Leadville in 1937. Colligan died at Buffalo on September 11, 1941, at the age of 74.

DAVID S. DIGERNESS COLLECTION

OPPOSITE: Buffalo was a town along the DSP&P that catered to the passenger business, as evidenced by the store fronts next to the main line. The year was 1898, when the Old South Park made extra provision for the patron with the roofed-over waiting room shelter added on to the depot. A spur track, used for lumber loading, ran beside the river.

A DSP&P MIXED TRAIN, consisting of 12 cars, paused beside the North Fork of the South Platte while W. H. Jackson exposed this view. The train was heading west, between Buffalo and Pine.

DAVID S. DIGERNESS COLLECTION

W. H. JACKSON PHOTO
DAVID S. DIGERNESS COLLECTION

CHARLES WEITFLE PHOTO
DAVID S. DIGERNESS COLLECTION

A MIXED TRAIN in Platte Canyon, pulled by a Mason bogie, was headed for Denver when it stopped on a curve for this stereoscopic view of 1879. The Denver, South Park & Pacific still was pushing its railhead westward at this time — its immediate goals: the rich mining districts of the southern Mineral Belt and the silver San Juans of southwest Colorado.

JOSEPH COLLIER PHOTO
DAVID S. DIGERNESS COLLECTION

DEER CREEK WATER TANK, at Milepost 48, was accessible only by the Denver, South Park & Pacific Railroad. This tank was replaced before 1890 with a larger tank at Crossons, to the east. This location is just west of Pine Grove.

DAVID S. DIGERNESS COLLECTION

PINE GROVE, COLORADO, was viewed from a hillside in 1924 looking toward the east. Railroad activity was at a temporary standstill, although rolling stock had been "spotted" at various locations throughout the yards. The Pine Grove depot is located within the wye in the lower right corner of the picture. A large trestle across from the depot fueled the locomotive with coal.

DAVID S. DIGERNESS COLLECTION

DEER CREEK, PLATTE CAÑON.

H. H. BUCKWALTER PHOTO
STATE HISTORICAL SOCIETY OF COLORADO

BUCKWALTER, a photographer and railfan, took this photo in the 1890's at Pine. The passenger car to the left was UPD&G 191 and the car to the right was DL&G 59. They were spotted on the ramp track leading to the top of the coal chute.

JOSEPH COLLIER PHOTO
DAVID S. DIGERNESS COLLECTION

THIS VIEW of Deer Creek Trestle was taken in the autumn of 1878, shortly after the construction of the DSP&P line to Webster. The Deer Creek water tank, shown on the right, was an important stop on this narrow-gauge route through rugged Platte Canyon. Arch Rock — on the left — was one of the numerous outstanding landmarks along the Old South Park.

W. H. JACKSON PHOTO
DAVID S. DIGERNESS COLLECTION

BLASTED ROCK filled the river channel along stretches of the Denver, South Park & Pacific, where the canyon walls close in. This location was above Crosson's Station. During construction, men worked from ropes suspended over the river; and, in some places, drill stems used to support platforms were left sticking out of the rock wall.

DELL A. McCOY PHOTO

THE OLD RIGHT-OF-WAY of the DSP&P had long been abandoned when this color view of the Baileys depot was recorded on Ektachrome. Today, even the little depot has been removed from the bank of the North Fork of the South Platte River — in the name of "progress."

OPPOSITE, ABOVE and BELOW: The townsite of Webster is part of this panoramic view, along the route of the Denver, South Park & Pacific over Kenosha Pass. The author was looking to the northeast from the grade leading over the pass when he shot this color scene. Trains of the DSP&P once clattered by on shining rails laid on the long-abandoned ties shown here. From this point, the line looped around through Webster, and vanished out of sight on what became the highway route down the canyon. On the horizon, the snow-covered summit of Mount Evans rises to an elevation of 14,264 feet. In the scene below, it can be seen that C&S Number 65 sported deer antlers atop her headlight when she was photographed about 1905. Engine crews often decorated their charges in this manner during the period when engines were assigned to specific individuals.

DAVID S. DIGERNESS COLLECTION

DAVID S. DIGERNESS PHOTO

RICHARD B. JACKSON PHOTO
DAVID S. DIGERNESS COLLECTION

A WESTBOUND SPECIAL PASSENGER TRAIN had just arrived at Baileys, a small mountain hamlet beside the

JANUARY 6, 1887

Georgetown Courier.

Terms, $2.50 per Year.

The bill conferring municipal suffrage upon women in Kansas, has become a law.

Last Thursday's Hurricane.

One of the most terrible wind-storms ever known in the State visited Colorado last Thursday. At the foot of the mountains the velocity of the wind reached a hundred miles an hour. A great deal of damage was done in the valley towns, and three trains were blown from tracks.

The Leadville Express over the South Park road was blown from a bridge near Littleton and all of the passengers, numbering twelve, were more or less injured. The coach was picked from the rails bodily and thrown thirty feet to the dry bed of the creek below. The smoking-car followed and fell upon its side, unlike the coach, which turned a complete somersault with its occupants. The engine alone remained upon the track. Even its security was jeopardized, and in order to prevent it from toppling into the gulch the engineer propped

North Fork of the South Platte River, in 1932.

her steady against the wind with inverted railroad ties.

The confusion that prevailed was terrible. In the coach were confined four females and a half a score of men bleeding and prostrate, buried beneath the debris of broken seats. To add to the horror of the situation the stove overtoppled, and the work of devastation by fire then began. A lady, Mrs. Duncan, by name, was sitting by the stove when the accident occurred, and as the wreck struck the bed of the creek the hot coals fell upon her clothing, and in an instant she was in flames. Her husband promptly enveloped her in a coat at the same time snatching from her neck the blazing silk shawl which she was then wearing.

The first to extricate himself from the car was the news-agent, and he, by the aid of Conductor Harlow, hewed down the door and began to liberate the passengers. By this time the fire had been checked and the work of taking out the wounded began. As fast as liberated they were removed to the ranch building close by, where Mr. Francis and household treated their wounds. Those injured and so cared for were as follows:

William Duncan, of Pine Grove, rib broken and otherwise injured in the back. Mrs. Lizzie Duncan, wife of the above, scalp wounds and injured breast. J. L. Oates, cattleman, Denver, face cut. H. B. Tuttle cattleman, Denver, arm broken and right eye severed from the head and a gash across the forehead. Mary Allen, housemaid, Denver, head hurt. M. F. Grove, paperhanger, Leadville, bruised about the loins. Mrs. Alice Klett, Como, hand cut. R. W. Wilson, Denver, manager of Great West Mine, of Alma, severe injuries in the head and a strained back. Mrs. Mary L. Jones, Como, milliner, injured internally, sprained in the right side and below the chest. David Fulton, Como, injured internally. J. W. Evans, mining engineer of Alma, contusion of hip and back. B. M. Morrison, Denver, postal clerk, back bruised, hand sprained. C. A. Scott, brakeman, back sprained and hand bruised.

After the injured passengers had been liberated the car again caught fire, and despite the efforts of the train men was almost entirely destroyed. The roaring of the flames and the fury of the tempest gave a dramatic effect to the scene which was heightened when the long grass in the gulch caught fire and swept to either side in great blazing sheets.

The Durango and Silverton Express, was blown from the track near Colorado Springs. The passengers escaped injury, but several of the trainmen were slightly hurt. One of the express cars caught fire and was totally destroyed.

Twenty-five empty freight cars were also blown from the track at Colorado Springs.

The Leadville Express to Denver was lifted from the track and hurled down an embankment, near Como, and the following persons injured:

Conducter McPherson, left leg slightly cut and bruised. Frank C. Ewing, of Denver, internal injuries in left side. G. Bryant, of Denver, shoulder and face bruised; also injured in left side. E.L. Purnell, of Leadville, sprained ankle. T. McDonald, employe, living at Kokomo, cut in back of his head.

DAVID S. DIGERNESS PHOTO

THE MINERAL SAMPLE on the left, mined in Lake County, is sibnite in barite with traces of Wulfenite and mimetite. In the lower right is a sample of galena and sphalerite with quartz. The pyritohedral crystals shown in the top right were mined in Eagle County.

MINERALOGY OF COLORADO

Sent prepaid on receipt of 25c. Address

J. S. RANDALL, Georgetown, Col.

J. S. RANDALL,

Dealer in Minerals

GEORGETOWN, COLORADO.

Collections containing from 50 *to* 100 *Minerals for sale,*

From $3.00 to $10.00.

P. L. RICE, M. D.,

OFFICE ROOM 4, FISH BLOCK,

Georgetown, Col.

OPPOSITE: Sun-bleached ties and the eroded grade of the Old South Park are all that remain of the narrow-gauge railway at the summit of Kenosha Pass. The scent of fall in the air rises from the distant aspen trees, where once it was possible to smell the acrid aroma of coal smoke blasted out of the stacks of little Mason bogies. Beyond is the vast expanse of South Park.

LOCOMOTIVE NUMBER 537 was leased to the Colorado & Southern by the CB&Q (today's Burlington Northern) in 1930, and lasted long enough to drag the last freight train out of the Foothills on April 11, 1937. Shown here at Baileys, it is easy to see that this 2-8-0 was an outside-frame engine.

OTTO C. PERRY PHOTO
DAVID S. DIGERNESS COLLECTION

DAVID S. DIGERNESS PHOTO

DAVID S. DIGERNESS COLLECTION

AT KLINE, COLORADO, the elevation was listed at 8,337 feet on the sign over the waiting house. This location, shown in 1905, was later known as Singleton, at milepost 62.5 on the DSP&P. A small private resort still exists here above Baileys, a few miles below Grant.

OPPOSITE, ABOVE and BELOW: The 537 was at Baileys in August of 1932 with a westbound seven-car special. The 537 originally ran on the Deadwood Central and was built by the Baldwin works in 1896. A short three-car spur (below) served the little resort of Glenisle, situated beside the North Fork of the South Platte. Notice the small, windowless waiting-room station next to the mainline of the DSP&P, just to the left of the boxcar, which afforded the passengers shelter. This is just above Baileys.

H. D. SMITH PHOTO
DAVID S. DIGERNESS COLLECTION

RICHARD B. JACKSON PHOTO
DAVID S. DIGERNESS COLLECTION

RICHARD B. JACKSON PHOTO
DAVID S. DIGERNESS COLLECTION

THE SPECIAL EXCURSION TRAIN pictured above and below traveled through the misty Foothills of the Front Range — above Baileys — on its way from Denver to South Park in 1932. The locomotive was Number 537, on lease from the CB&Q.

RICHARD B. JACKSON PHOTO
DAVID S. DIGERNESS COLLECTION

ALEX MARTIN PHOTO
RICHARD RONZIO COLLECTION

THE TOWN OF GRANT, shown here as it appeared in 1880, was located beside the North Fork of the South Platte River. In this view, the photographer was looking east. A coal bunker, section house, depot, water tank and wye served the DSP&P at this station.

CHARLES WEITFLE PHOTO
DAVID S. DIGERNESS COLLECTION

WEBSTER STATION was the end-of-track in 1879, when this train had just terminated its run beside the depot. On the siding, goods from a boxcar were being unloaded onto the freight dock.

JAMES W. NUTT PHOTO
DAVID S. DIGERNESS PHOTO

CASSELLS was another of the many resorts along the DSP&P beside the North Fork of the South Platte River. It was listed at milepost 64.5, approximately 10 miles above Baileys. Two ladies used umbrellas as shade against the bright sunlight, and to the right, one man has his shotgun at hand. The maids, cook, manager and children all came outdoors for this occasion in 1885.

THE LAST DAYS of railway operation on the Old South Park found the Grant depot in a sad state of affairs. The station sign on the left side of the building read, "Elevation 8,555 Feet."

DAVID S. DIGERNESS COLLECTION

H. E. HIGH PHOTO
DAVID S. DIGERNESS COLLECTION

THE CHARCOAL OVENS at Webster served The Mineral Belt's smelting industry, including the Boston & Colorado smelting works at Black Hawk in Gilpin County. The ovens—located on the Denver, South Park & Pacific—were not being used at the time this photograph was taken, but were very active in the 1880's.

WEBSTER CAN BE SEEN from the grade of the South Park line in this view of 1879, taken from a position similar to the color view above.

CHARLES WEITFLE PHOTO
DAVID S. DIGERNESS COLLECTION

W. H. JACKSON PHOTO
RICHARD RONZIO COLLECTION

THIS "PHOTOGRAPHIC SPECIAL" — shown just west of Webster station — was provided for W. H. Jackson's use by the Denver, South Park & Pacific Railroad. Jackson was on his return trip to Denver when he shot this picture in 1886. In the background, a freight train can be seen descending the grade from Kenosha Pass.

Chapter **VII**

Kenosha Pass to Granite

BEFORE WE PROCEED ANY FURTHER, let's go back a number of years to the late spring of 1860. At this time, the surface dirt of California Gulch was panned by a party of Gilpin County prospectors, and they found it to be rich in gold beyond expectation. One member of the party yelled, "Gold! . . . I have all of California in my pan!" and the gulch became known as *California Gulch*.

Parts of the gulch were fabulously rich. The original discoverers — Currier, Lee, Rafferty, Slater and Stevens — secured claims 200 feet up and down the stream, and those who came after were allowed 100 feet, until the entire gulch — more than five miles in length — was taken. A miner who was there at the time told Jesse, "The gulch was quite a sight to be seen. There was almost a continuous scene of activity bordering the stream and skirting the bases of the hills of the gulch. The water supply was limited, so it was used over and over again by each successive claim owner, and by the time it reached the lower part of the gulch, was the consistency of liquid mud. Some claims yielded over a thousand dollars a day."

Before the summer was over, more than 5,000 men had come into the area, and the camp was the most productive and prosperous in the mountains. In the meantime, Oro City was established at the upper end of the gulch and was the center of trade and traffic. The usual miners' laws were in force, and the locality had much the same characteristics as other early mining settlements. There were the usual number of gambling houses, saloons and parlor houses, and when night came, many a miner lost the rich earnings of the day among them. This early camp saw its best day in 1861.

The working season of California Gulch was quite limited, for the elevation is over two miles above sea level. The winters are long and linger far into spring, preventing most work from being done outside the summer months. Every year, when the deep snow came to stay, most — if not all — were forced to leave the area and wait out the winter season at lower elevations.

California Gulch continued as a mining camp for a number of years, until the most accessible ground had been worked over and the gold production dropped drastically. During the summer of 1864, a good number of the miners, who were so plentiful in the gulch, started moving on to newer diggings, and by 1866, almost the entire population had left the area.

The most reliable data at hand indicates a yield of $1-million in gold the first summer and a total yield of nearly $4-million by the close of 1865.

The California Gulch area was pretty dead by 1868, when the Printer Boy lode was discovered. This rich mine caused quite an excitement, and much prospecting was done at the time, several gold lodes being located in the area. Nearby, a *new* Oro City formed, which was two and a half miles farther up California Gulch from the other, or *old* Oro City, where a number of the log buildings remained as landmarks of those early days of the first settlement. It had been located at a place where the hills break away toward the present site of Leadville. The new Oro City, Jesse observed, continued to be a place of importance, even though the Printer Boy was the only really profitable mine worked. This mine showed an immense body of soft quartz that was extra rich in gold, and it was not long until a 25-stamp mill was erected. Up to 1875, the mine yield exceeded $250,000 from work carried on for six years during just the summer months.

Previous to this and other gold-lode discoveries in California Gulch, profitable mining of a similar character had taken place some 17 miles down the Arkansas River Valley near the town of Granite. This began in 1866 and continued for about three years before the surface pockets were exhausted and work stopped.

In the years succeeding 1871, some attempts at ore treatment were made at Granite. Also, a few discoveries were made 30 miles north of California Gulch, near the headwaters of the Tennessee Fork of the Arkansas River. One of these — called the Homestake — was of great size and rich in lead, but poor in silver. However, as a result of ore shipments made from the Homestake to the Golden Smelter Works (located at Golden), and with the known presence of lead-bearing silver lodes (or veins) among the hills of the Park, or Mosquito, Range, along the east side of the valley of the Arkansas River — a smelter was constructed near the confluence of California Gulch and the Arkansas River. The place was named Malta, and the Malta Smelting Works originally was fired up and began to conduct business in the fall of 1876, but shut down until it was modified a year and a half later.

W. H. Stevens, a miner from South Park, accompanied by

RICHARD H. KINDIG PHOTO
DAVID S. DIGERNESS COLLECTION

THE STEEP GRADES on the east side of Kenosha Pass once again felt the drivers of C&S 73 as she toiled up the grade with a dismantling train. This was taken on July 14, 1938. The town of Webster can be seen in the valley below.

SAFETY RAIL WAS ADDED to the DSP&P line over Kenosha Pass after a train wreck tore up the track above Webster in the 1880's. The original light rail had been discarded beside the grade. Far below, the stage road to Leadville wound its way toward the pass.

ALEX MARTIN PHOTO
DAVID S. DIGERNESS COLLECTION

A. B. Wood, purchased the California Gulch placer claims in 1874; and, for several years, worked on the construction of a 12-mile ditch to divert water to the gulch. The plan was to bring in a sufficient water supply to permit hydraulic mining for the cheap handling of unworked gravel banks which bordered the gulch. The upper Arkansas River was the source of supply, and several summers were required to complete this undertaking, the enterprise being known as the Oro Ditch & Fluming Company. The first full season of work was put in on the placers in the summer of 1878.

California Gulch had been settled for 15 years, and had been worked over and over again, until it had yielded millions of dollars in gold; whereupon, it was believed that the mineral wealth had been exhausted and the area was all but abandoned. Yet, close by, a seemingly inexhaustible treasury

W. H. JACKSON PHOTO
DAVID S. DIGERNESS COLLECTION

of precious metals lay embedded, unknown and untouched until the great hydraulic mining period got under way in 1878, and it was still going strong in 1879. All the time that gulch mining had been going on in this area, the miners had to contend with the inconvenience of the many large and very heavy boulders which they had to handle many times, moving them over and over again in the creek. None of them knew, or stopped to investigate, the mineral content of the boulders that caused them so much inconvenience. Two men — by the names of Stevens and Wood — discovered the boulders to be carbonate of lead which carried silver, but they did not make their discovery known until they were fairly certain of government titles to nine claims. These were taken up lengthways along what they considered to be the crest of the lodes, and each claim took up an area of approximately 1,500 by 300 feet. The territory crossed California Gulch and extended high up on the hills. Commencing at the south, the

ONE OF W. H. JACKSON'S special Denver, South Park & Pacific trains traversed the eastern slope of Kenosha divide in 1887. The special — typical of trains of its day — was polished to a mirror finish. The locomotive is Number 42, one of the DSP&P's trim little 2-6-6T Mason bogies, originally named the "Ten-Mile."

OVERLEAF: William Henry Jackson aimed his giant 20x24-inch view camera eastward toward the Great Plains on his DSP&P expedition of 1887 to capture this view of Webster — down in the valley, far below the railway's upper-level grade. Charcoal kilns and commercial buildings made up the tiny settlement. Westbound Number 42 and coach 9 (used for a darkroom) were near Jackson's camera, on the eastern approach to Kenosha Pass — "running extra."

W. H. JACKSON PHOTO
DAVID S. DIGERNESS COLLECTION

RICHARD H. KINDIG PHOTO
DAVID S. DIGERNESS COLLECTION

COLORADO & SOUTHERN Number 73 reaching the summit of Kenosha Pass in 1938.

OTTO C. PERRY PHOTO
DAVID S. DIGERNESS COLLECTION

ONE OF THE LAST stockcar train movements took place April 11, 1937, with cars gathered from the line as far away as Leadville. In this photograph, the double-headed train started to descend the east side of Kenosha Pass, heading toward Denver. Chicago, Burlington & Quincy Number 537 is on the lead, with Colorado & Southern Number 72 following. These 47 empties were gathered from towns along the line from Climax to Como, due to the abandonment.

LITTLE NARROW-GAUGE C&S 73 was photographed mounting Kenosha Pass July 14, 1938 with a dismantling train. She was westbound out of Webster near the top of the pass when this portrait was taken.

RICHARD H. KINDIG PHOTO
DAVID S. DIGERNESS COLLECTION

CHARLES WEITFLE PHOTO
DAVID S. DIGERNESS COLLECTION

A CONSTRUCTION TRAIN was heading upgrade on Kenosha divide, west of Webster — rolling over new rail and crossties. Kenosha Pass can be seen just over the engine cab in this 1879 print.

names of the principal locations were the Dome, Rock, Stone, Lime, Bull's Eye and Iron. The ore was first found in place in a lode in the Rock claim, where it was over 10 feet thick. It was rich in lead, but carried only a small amount of silver.

Other miners, such as one Mr. Durham, claimed to have discovered carbonates prior to this, when he dug a shaft down into the Oro La Plata lode. Still another couple of miners — Maurice Hayes and his brother, who for several years ran an assay office at Oro City — had been prospecting extensively and found some of the earliest carbonate of lead locations in the district.

The first strike considered to be exceptional was made by the Gallagher brothers on the same hill as the location of the Iron lode, but nearer Stray Horse Gulch. This claim, known as the Camp Bird, was discovered late in the fall of 1876.

The St. Louis Smelting & Refining Company, located in the town of Alma for quite a few years prior to 1877, purchased ores in both Park and Lake Counties. The prospective production of the Rock, Camp Bird, and other claims caused the company to establish a sampling works on the present site of Leadville. Shortly thereafter, the St. Louis Smelting Company erected a smelter in Leadville and had a blast furnace in operation by October of 1877. To insure an ore supply, a contract was made with Stevens & Wood to furnish 1,000 tons of ore from the Rock mine. Before the entire amount of ore had been delivered, such developments had taken place and so many discoveries had been made that the sampling and smelting works were having difficulty handling the large volume. The smelter was named the Harrison Reduction Works, after the president of the company, Edwin F. Harrison. About this time, A. B. Wood sold his interest in the nine claims to L. Z. Leiter, of the great Chicago dry goods firm of Field, Leiter & Company, for $40,000. The true value of the Iron lode (or claim) was then unknown and the low-grade Rock mine was considered the best of the lot. In the fall of 1877, the former claim began to be worked extensively

W. G. CHAMBERLAIN PHOTO
STATE HISTORICAL SOCIETY OF COLORADO

KENOSHA PASS had this small depot with a passing track at 9,991 feet in elevation. Game abounded in the area, as evidenced by the deer carcass mounted over the railing. This photo was taken in 1879, shortly after rail reached this point.

JOSEPH COLLIER PHOTO
STATE HISTORICAL SOCIETY OF COLORADO

A TRAINLOAD OF TIES, headed by a Congdon stack Mason bogie, stopped for the photographer near the top of Kenosha Pass in 1879. This photograph, with a view of South Park, was the result of considerable effort on the part of the photographer. The large camera had to be removed from the train, placed in position and carefully loaded with a sensitized emulsion-covered glass plate negative. The exposure often took a minute or more. The cases for the camera equipment appear at the right of the photo.

A CONGDON STACK MASON BOGIE locomotive was stopped on the DSP&P grade overlooking South Park. This photo was taken from the west side of the grade climbing Kenosha Pass around 1879. At the time, the DSP&P was still constructing mainline track across South Park, near Trout Creek Pass.

CHARLES WEITFLE PHOTO
STATE HISTORICAL SOCIETY OF COLORADO

W. H. JACKSON PHOTO
DAVID S. DIGERNESS COLLECTION

THE WIDE-OPEN MEADOWLAND of South Park stretches out beyond this special W. H. Jackson train in 1886. Eastbound engine 42 and waycar 72 are almost at the top of Kenosha Pass — DSP&P's eastern gateway to this great natural mountain park.

and proved to be of immense value, even outdoing the top-rated mine, the Camp Bird. The Iron mine, more than any other property, gave this mining district its first great fame. It held the lead until the time of the Fryer Hill developments in the fall and winter of 1878.

In the fall of 1877, several hundred men were in the district prospecting or sinking prospecting shafts, and the amount of mining done was considerable. The ore shipments were steadily increasing in volume and a vast amount of low-grade mineral was mined, for which there was no market. It gradually became more and more evident that the mineral wealth of the California Mining District was immense. Mines were being developed on every hand, notwithstanding the fact that it was necessary to team ore 120 miles by wagon road to the plains, and then send it by rail nearly a thousand miles further to St. Louis or Omaha. Shipments to the latter place were made by Berdell & Witherill, who had established a sampling mill in the district late in 1877. To meet the growing demands of travel, McClellan & Spotswood extended their Denver & South Park stage lines to Leadville, and several ore transportation and freight lines were established between Leadville and the DSP&P's rail-head. The best paying mines at that time were the Iron, Camp Bird, Carbonate and the Long and Derry.

Leadville's population increased steadily during that winter and the following spring; and, during this time, a school and several churches, as well as a weekly newspaper, were established. In March of 1878, the first important transfer of mining property was made. Representatives of the St. Louis Smelting Company, having become aware of the value and quantity of ore in the Camp Bird claims — through purchases at their smelters — bought it and the adjoining Pine, Keystone and Charleston claims for $225,000. The Pine and Charleston were staked in late fall of 1876, nearly the same time as the Camp Bird.

The fame of the *Carbonate Camp* had spread far and wide before the summer of 1878, and a flow of immigration had started that would continue for some time at an ever-increasing pace. The discoveries already made were extremely rich and productive; however, the development that gave Leadville

DAVID S. DIGERNESS PHOTO

THE SCENIC GRADE of the Old South Park — visible to the left of the aspen trees — descended Kenosha Pass to cross South Park. The Continental Divide flanks the north edge of the park, seen here during one of Colorado's sparkling Indian summers. A new mantle of snow dazzles the eye — and portends much more to come.

its great supremacy over any other place in The Mineral Belt — or anyplace else in the world at that time — was that which took place at Fryer Hill. Claims opened in the summer of 1878 were adequately explored by the following fall to give sufficient indication of their extreme extent and value.

Every day saw the volume of immigration of newcomers grow larger and larger with the arrival of wagon trains, stagecoaches and freight teams. There also were those that arrived on horseback and on foot. Among the fortune-hunters were all types of people, from rich to poor. Many hoped to "strike it rich" and were ambitious enough to do it. They came from everywhere and were of every profession — swelling the population of the Carbonate Camp almost beyond belief. Of course, the reason behind Jesse's trip to Leadville was the loss of so many people from the Georgetown area. These people also had come to this new *Eldorado* in the southern part of The Mineral Belt. He had to see it for himself. New blocks of buildings were built up in a single week. This was Leadville before it grew into a city of metropolitan proportions.

The district growth was greatly accentuated and its fame was further enhanced by the wonderful development and production of the Fryer Hill mines in the summer and fall of 1878. These ore deposits were so rich and plentiful and so easily mined that they surpassed anything previously discovered in the state or territory, and virtually anything outside

DAVID S. DIGERNESS COLLECTION

A SIX-CAR FREIGHT TRAIN was ascending Kenosha Pass from South Park in this view of 1920, when the railroad was operated by the C&S. The highway crossed the pass using the higher grade. WILLIAM A. GIBSON PHOTO DAVID S. DIGERNESS COLLECTION

WILLIAM A. GIBSON PHOTO
DAVID S. DIGERNESS COLLECTION

OPPOSITE AND ABOVE: Taking on water at Jefferson in a snowstorm was a bit of a chore on the C&S. This was March 29, 1937, with locomotive Number 9 pulling the westbound mail train. A steam hose was fitted to the steam dome and used to thaw the frozen tank.

Denver and South Park Division.

DENVER TO LEADVILLE AND ST. ELMO.

Miles	STATIONS. ‡ Meals.	Elevation.	No. 401. Leadville Express.	No. 403. Leadville & Aspen Exp.
..	Denver Lv	5203	8 00 a.m.	8 30 p.m.
5	Mooreville	5313	8 22	8 52
7	Bear Creek Junction	5312	8 25	8 55
11	Littleton	5381	8 35	9 05
21	Platte Canon	5518	8 55	9 30
27	Deansbury	5877	9 18	9 55
29	South Platte	6112	9 29	10 04
32	Dome Rock	6227	9 37	10 13
35	Dawson's	6418	9 47	10 23
40	Buffalo	6651	10 03	10 45
42	Pine Grove	6766	10 16	10 58
48	Crosson's	7077	10 38	11 22
52	Estabrook Park	7578	10 56	11 40
55	Bailey's	7740	11 05	11 50
59	Slaght's	8043	11 18	12 05 a.m.
..	Cassell's	....	....	.....
66	Grant	8583	11 47	12 35
69	Webster	9005	11 58	12 52
76	Kenosha	10000	12 30 p.m.	1 29
81	Jefferson	9518	12 50	1 50
88	Como Lv	.....	1 35	2 25
93	Half Way	10557	2 05	2 55
98	Boreas	11498	2 25	3 25
103	Argentine	10608	2 50	3 50
110	Breckenridge	9555	3 19	4 20
116	Dickey	9510	3 41	4 45
118	Dillon ..... Ar	.....	4 15	....
124	Keystone ..... Ar	.....	4 35	....
126	Wheeler Lv	9702	4 22	5 30
132	Kokomo	10606	4 47	6 00
134	Robinson	10851	4 56	6 10
137	Climax	11325	5 09	6 25
138	Alicante	....	5 16	6 31
151	LEADVILLE ..... Ar	10213	6 00 p.m.	7 15 a.m.
..	DENVER Lv	5203	.....	8 30 p.m.
88	Como	9803	.....	2 30 a.m.
94	Red Hill	9548	.....	2 47
97	Hay Ranch	.....	.....	3 00
104	Garo's	9201	.....	3 20
113	Platte River	8962	.....	3 45
120	Hilltop	9153	.....	4 06
127	McGee's	8782	.....	4 35
131	Charcoal	8008	.....	4 50
132	Schwanders ..... Ar	7835	.....	4 58
137	Buena Vista ..... Ar	7957	.....	5 13
....	Schwanders Lv	7835	.....	5 28
142	Heywood Springs (Hortense)	.....	.....	6 02
149	Alpine	9175	.....	6 41
153	St. Elmo,	9998	.....	7 30
158	Hancock	10939	...	7 55
161	Alpine Tunnel	11624	.....	8 20
175	Pitkin	9108	.....	9 22
181	Ohio City	8475	.....	9 48
189	Parlin's	7837	.....	10 21
201	GUNNISON ..... Ar	7582	.....	11 00
216	Baldwin	.....	....	
218	Baldwin Mine	.....	....	

Denver and South Park Division.

Between LEADVILLE, ST. ELMO and DENVER.

Miles.	STATIONS. † Meals.	Elevation.	No. 402. Leadville Express.	No. 404. Leadville & Aspen Exp.
...	Leadville Lv	10213	8 00 am.	8 45 p.m
13	Alicante	.....	8 47	9 35
14	Climax	11325	8 54	9 45
17	Robinson	10851	9 06	9 58
19	Kokomo	10606	9 14	10 05
25	Wheeler	9702	9 38	10 30
35	Dickey	9510	10 16	11 05
....	Keystone Lv	9160	8 00	
....	Dillon	8847	8 20	
41	Breckenridge	9555	10 37	11 30
48	Argentine	10608	11 07	12 07 a.m
53	Boreas	11498	11 40	12 40
58	Half Way	10557	12 05 m	1 05
63	Como Lv	.. ..	12 45 †	1 40
70	Jefferson	9518	1 1[illegible]	2 10
75	Kenosha	10000	1 30	2 30
81	Webster	9005	2 02	3 02
85	Grant	8583	2 15	3 15
86	Cassell's	.. ..	.....	.....
92	Slaght's	8043	2 41	3 41
96	Bailey's	7740	2 58	3 58
99	Estabrook Park	7578	3 08	4 08
103	Crosson's	7077	3 25	4 25
109	Pine Grove	6766	3 44	4 44
111	Buffalo	6651	3 54	4 54
116	Dawson's	6418	4 10	5 10
119	Dome Rock	6227	4 23	5 23
121	South Platte	6112	4 32	5 32
124	Deansbury	5877	4 41	5 41
130	Platte Canon	5518	5 03	6 03
140	Littleton	5381	5 29	6 29
143	Bear Creek Junction	5312	5 38	6 38
144	Mooreville	5313	5 40	6 40
151	Denver ..... Ar	5203	6 00 p.m	7 00 a.m
...	Baldwin Mine	.....	....	....
2	Baldwin	.....	....	....
17	Gunnison Lv	7582	....	4 30 p.m
29	Parlins	7377	....	5 10
37	Ohio City	8435	....	5 43
42	Pitkin	9108	....	6 10
57	Alpine Tunnel	11624	....	7 20
60	Hancock	10939	....	7 40
65	St. Elmo Lv	9998	....	8 30
70	Alpine	9175	....	8 50
77	Heywood Sp., (Hortense)	.....	....	9 24
86	Schwanders	7835	....	9 58
90	Buena Vista Lv	7957	....	10 13
86	Schwanders	7835	....	10 28
87	Charcoal	8008	....	10 35
92	McGee's	8782	....	10 58
99	Hilltop	9153	....	11 27
100	Platte River	8962	....	11 51
114	Garo's	9201	....	12 21 a.m
121	Hay Ranch	.....	....	12 44
130	Como Lv	9803	....	1 40
218	Denver ..... Ar	5203	....	7 00 a.m

DAVID S. DIGERNESS PHOTO

of Colorado except the Comstock lode in Nevada. Meanwhile, the Iron, Argentine and other great mines kept on producing as heavily as ever, and every month saw fresh additions to the many paying mines. The people — now having ample shelter — stayed the year-'round. The icy winds of winter came, but failed to hinder the continuing caravans of wagons headed for the new land of promise. When the snow got too deep for wheeled vehicles, runners were used on the wagons. No matter how distant the land or locality, all roads seemed to lead to Leadville.

The railways directed their attention to the scene and headed their roads toward nearly two-mile high Leadville — the *Cloud City*. Wagon roads were jammed with long lines of teams bringing freight and immigrants into the area. Stage lines increased their passenger capacity; and yet, travel increased faster than all the modes of transportation available to the Carbonate Camp could accomodate. Some of the heaviest travel from the plains was by way of Colorado Springs and Manitou over Ute Pass, where long wagon trains were constantly ascending and descending the mountain defile — amid almost perpendicular walls of rock. Beautiful scenery,

THE JEFFERSON DEPOT once again has come back to life, now occupied by a real estate office (in 1977). Many of the original houses of Jefferson still exist, even in the harsh climate of South Park. C&S locomotives have not "whistled for the board" here for nearly 40 years.

OPPOSITE, ABOVE and BELOW: The "Old South Park" was reasonably well-maintained by the Colorado & Southern, even in later years—as evidenced by this Jefferson station scene, taken in 1929. The Jefferson depot had changed little since its construction. In this photograph, C&S passenger train 70 was pausing on its regular daily run between Denver and Leadville. On the point of the train was locomotive Number 10, a Class B-3-C 2-6-0 of 1884 vintage. The C&S listed Jefferson at Milepost 81, at an elevation of 9,730 feet above sea level. The back side of the Jefferson depot, as it appeared in 1942 (below), shows extensive remodeling around the doorway. The rail was gone from the roadbed and the little depot was in a state of decay.

RICHARD B. JACKSON PHOTO
DAVID S. DIGERNESS COLLECTION

CARROLL WEIS PHOTO
KLAAS GUNNINK COLLECTION

M. C. POOR PHOTO
DAVID S. DIGERNESS COLLECTION

THE SOUTH PARK STATION of Jefferson — with its neat little wooden-frame depot — was all but forgotten by the time this photograph was taken in 1937. The main-line extended in a southwesterly direction across South Park, toward the Park Range en route to Leadville.

consisting of glens, grottoes, streams and waterfalls, greeted travelers on their way to and from Leadville via South Park.

More than a dozen sawmills were turning out lumber night and day, and still could not meet the demands of the Carbonate Camp's construction boom. The streets of the entire camp were one whirl of excitement, and each successive month only added to the animation that pervaded all industries and every branch of trade. Men who had made fortunes from working or selling their mines turned about and bought up prospects, or newly developed properties, at what would have seemed outrageous prices not very long ago. Many newcomers came in the middle of the winter the hard way. The snows and suffering attending a trip over the great mountain ranges held no fears for those who had caught the Leadville fever and headed with the stampede. Fearing that by delay, they would lose their great opportunity for striking it rich, an average of 100 people a day arrived in Leadville much of the time from January to May.

Hard-rock mining in this lusty town that proudly called itself the *Magic City* was not the only money-making business. Fortunes were made in real estate, and several successive advances in the values of town lots were sources of large windfall profits to many a property holder and speculator. Building sites that could be purchased for $25 in the summer of 1878 were selling for several thousand dollars the following winter, and a single week saw real estate values go up four times in price.

Chestnut Street (later renamed First Street), for a time, was the main business thoroughfare, but Harrison Avenue shortly took over this honor. Land values were running from $100 to over $250 per foot of frontage on these streets.

Jesse noted that three daily newspapers were in business in

M. C. POOR PHOTO
DAVID S. DIGERNESS COLLECTION

JEFFERSON TANK and a harp switchstand were the subject of this photograph taken in 1937. The Tarryall Mountains appear in the background, while in the distance (to the right), the C&S grade over Kenosha Pass can be discerned if one looks closely.

Leadville. Every trade and calling proved a source of revenue, and merchants, lumbermen, mechanics and professional men had all the business they could attend to. One business helped another, and all helped the freighter. The transportation of freight to and from the DSP&P railway terminal (end-of-track) was a somewhat fascinating spectacle to watch — very similar to Georgetown's situation when the Colorado Central terminal was still down valley from it. The DSP&P already was at a point where Leadville was giving it a tremendous amount of business. The Colorado Springs freight lines alone had 2,000 animals engaged in the transportation business, and yet lacked men and mules to such a degree that they could not handle all the business that was available to them.

By 1879, much construction was in progress and all of it resulted in the largest mountain town in the state. Business houses — large and small — were there by the hundred, and there were nearly a hundred licensed saloons and a dozen gambling and parlor houses running full blast, both day and night. Along the main business street it was difficult to make one's way by day, and even more difficult just before nightfall, when the miners came into town in swarms. Many bands could be seen drumming-up audiences for the theaters and variety shows, as well as for the scores of saloons and numerous gambling houses. Many of these places did a tremendous business, for those who make money quickly and easily, generally spend it just as quickly and easily.

In one way or another, nearly every business or professional man, clerk, saloonkeeper, blacksmith and mill employee in Leadville was engaged in mining or prospecting. This was often done by furnishing grubstakes to prospectors in need of

DAVID S. DIGERNESS COLLECTION

THE BATTLE WAS LOST, as this view of a C&S locomotive shows — with the engine lodged tight in the snow. This may have been a helper locomotive being turned on a wye, since the tender is in view. The broad expanse of South Park allowed formidable winter blizzards to develop, which often blocked the Colorado & Southern mainline. However, these blinding "white-outs" usually were short-lived — and the sun would burst forth in dazzling splendor in the middle of an unbelievably deep-azure sky — as if there had been no storm at all.

C. O. DAVID PHOTO
DAVID S. DIGERNESS COLLECTION

BURIED IN DEEP SNOW, this South Park train had stalled a few-hundred feet from Como during a winter blizzard of 1904. Even with three engines, the train was unable to forge through the field to its destination. Notice the shovel left on the tender by the fireman of the lead locomotive.

DAVID S. DIGERNESS PHOTO

ONLY A FEW ROTTEN TIES are left to mark the grade of the Old South Park, where narrow-gauge trains once approached Como. The company's roundhouse, depot and hotel still stand in this all but forgotten railroad town. A wooden water tank once stood near this spot, about a mile from the DSP&P depot.

supplies, which consisted of food and supplies, and prospecting and mining tools. In return, the prospectors gave the grubstakers an interest in any discovery they might make. Sometimes, even wages were furnished to miners who would sink a shaft, while the backer of the enterprise continued to follow his own trade.

Leadville's population was just under 1,500 by June of 1878; and, by the following October, it had doubled. A census taken in January of 1879 showed 5,040 residents, not counting all of those living just outside the town limits. After that time, and up to Jesse's visit to Leadville, the town grew beyond anyone's expectation, and the most moderate estimate of the population of the Carbonate Camp ranged from 12,000 to 15,000.

Meanwhile, as mentioned previously, H. A. W. Tabor had set up one of Leadville's first general stores in the summer of 1877, and he grubstaked many a prospector. It was mainly from grubstakes that Tabor accumulated his substantial wealth during Leadville's great silver boom. Tabor's store performed a special service to the early Leadville Mining District in providing postal services, as well as supplying groceries, tools for prospectors, hardware of all styles, drygoods, etc. Consequently, Tabor's general store was a gathering place and headquarters for the area's people where they could buy supplies and even indulge in a game of cards and drink some of Tabor's fine whiskey. Here, they could pick up their mail and discuss the latest news.

Tabor's general store turned out to be the stepping stone

M. C. POOR PHOTO
DAVID S. DIGERNESS COLLECTION

THE ABANDONED King Coal mine was located approximately three miles south of Como, in South Park. Large amounts of low-grade lignite coal were mined here. This coal was used for fuel on the South Park line.

M. C. POOR PHOTO
DAVID S. DIGERNESS COLLECTION

THIS FAN HOUSE — now in ruins — was built at the King Coal mine in 1880. The King Coal mine, located on the DSP&P near Como, had a major coal-dust explosion in 1893, which killed 27 miners. The railway ran a special train to convey the dead to Como.

M. C. POOR PHOTO
DAVID S. DIGERNESS COLLECTION

MINE DUMPS and building foundations were all that remained of the town of King in 1941. The town, located in South Park, was named for C. W. King, a bookkeeper for the South Park Coal Company. The King Coal mine, which supported this settlement, was on a branch-line of the DSP&P, slightly more than three miles from Coal Branch Junction — near Como.

"STRIKING IT RICH" was a dream that drove men to do almost insane things—such as heading out across South Park by buggy or wagon in the midst of snowstorms. An average of 100 people stricken with "gold fever" arrived in Leadville each day from January through May during the height of the mining boom.

DAVID S. DIGERNESS COLLECTION

to fantastic riches for him. He furnished a grubstake in the month of May in 1878, to a couple of prospectors by the names of August Riche and George T. Hook. The grubstake consisted of some mining tools and groceries for a one-third interest in whatever they might discover.

While the Leadville mining camp was excited over the Iron, Camp Bird and Carbonate mines, George H. Fryer sunk a shaft on a hill north of Stray Horse Gulch. He found low-grade carbonate on April 4, 1878, and called his claim the New Discovery. This was the beginning of what became the best part of the best-paying mine in Colorado at the time Jesse visited Leadville in 1879.

Luck was with Tabor's two grubstaked prospectors. When they went up Fryer Hill to prospect, they first stopped to rest — and that was where they decided to dig. After sinking their hole, August Rische and George T. Hook discovered a great mineral deposit. The ore strata was unusually close to the surface at that point and required little work to reach it. The first wagon load of ore proved to be quite valuable and netted the owners around $300.

Rische and Hook started a drift from the shaft hole along the direction of their vein, and it grew larger and larger until it reached the point where it was several feet thick. The discoverers then were sure they had struck a vein that would

DAVID S. DIGERNESS PHOTO

COMO, A FORMER JUNCTION and division point of the DSP&P, is included in this autumn vista, looking across the northern edge of South Park toward the snowy Park Range. The old railway hotel, depot and stone roundhouse still stand here as mute evidence of the importance this town once had as a narrow-gauge center.

make them a fortune. The claim was called the Little Pittsburg after George Hook's old home town in Pennsylvania. At this time, the only other mine anywhere nearby was a claim known as the Union — farther east — where a deep shaft had been sunk; and, as soon as the news of the Little Pittsburg ore sales spread, there was a general rush to this virtually unknown part of the district. In June, a work force was hired and mining was then pushed at a fast rate of speed. In the last half of July in 1878, the Little Pittsburg and New Discovery each were yielding 75 tons of ore per week. The Little Pittsburg's ore milled from 90 to over 200 ounces of silver per ton. The New Discovery's ore did not average much over 50 ounces per ton, but even at that figure it was quite profitable. The locality of this activity was called Fryer Hill after the first discovery in 1879. During the summer of 1878, men were sinking shafts on all parts of the hill, and ore was found in the Little Chief, Carboniferous and Chrysolite claims.

The Little Pittsburg was paying thousands of dollars per week during the month of August and in September was the leading producer of the district. It was during September that George Hook sold his interest in the Little Pittsburg to his partners, Rische and Tabor, for a total of $98,000. The buyers made the purchase price of the mine in just three weeks of operation.

Before Rische and Hook started their prospecting on the hill, another shaft had been started nearby. When they filed their location papers on the Little Pittsburg at the land office,

DAVID S. DIGERNESS PHOTO

MASSIVE PILES of gravel and rock were left behind when this abandoned dredge made its way across the vast reaches of South Park — forever scarring the land. A large roadgrader sitting near the pond provides some idea of the size of the dredging boat beyond it. The ridge of peaks to the northwest is part of the Mosquito Range, as well as part of the Continental Divide. Beyond these high peaks is the great mining center of Leadville.

they took in the then-barren shaft known as the Winnemuck. Before that time, four men named Bissell, Foss, Trimble and Hunter had purchased the Winnemuck. They did not strike mineral until the shaft was down to a depth of 150 feet — at which point they had from 10 to 30 feet of mineral, but it was claimed by Rische and Tabor. Near the end of September, 1878, Bissell and his partners went over to the other side of the Little Pittsburg and bought out Borden's half interest in the New Discovery for $40,000. It so happened that they cleared that and much more in a few weeks work in the Winnemuck.

This purchase was a serious threat to Tabor, as the older title of the New Discovery might eventually take in the Little Pittsburg if the ore strata was decided to be a regular vein, or lode. Luckily, before the Winnemuck men could secure Fryer's interest, Senator Jerome B. Chaffee bought it for $50,000. The contents of the New Discovery was not then thought to be worth the total of $90,000, but was valuable for its position and older title. Before the senator left town, Tabor and Rische approached him — fearing that the older title of the New Discovery might imperil their claim to the Little Pittsburg — and bought the New Discovery for the handsome figure of $125,000.

The Leadville mines kept on improving faster than ever. Senator J. B. Chaffee and D. H. Moffat watched over the happenings and became confident and enthusiastic over the

RICHARD H. KINDIG PHOTO
DAVID S. DIGERNESS COLLECTION

THIS WAS THE END of operations over the Old South Park as the dismantling train worked across South Park near Jefferson on September 17, 1938.

DAVID S. DIGERNESS COLLECTION

DAVID S. DIGERNESS COLLECTION

THE BACKYARDS of Fairplay look out over the unsightly dredge piles and diggings of abandoned mining operations on the South Platte River. The town of Alma is located to the northwest, just this side of the distant peaks. Placer mining was extensive throughout this area. Historical native-stone monument in memory of the Tarryall Diggings—near Jefferson—in South Park.

OPPOSITE, BELOW: A Bucyrus dredge boat is shown here, afloat on its pond on the flood plain of the Blue River, not far above its confluence with the Swan River. During the late 1870's, placer operations were inaugurated by large dredging firms in the southern part of The Mineral Belt. This operation—north of the town of Breckenridge—belonged to the Colorado Gold Dredging Company. Huge buckets clawed away at the gold-bearing gravel (at the left) and deposited the valuable minerals inside the boat. Belts then carried the worthless gravel out behind into the dredge, where it was dumped onto ugly piles—in obvious disregard for the once-beautiful valley.

DAVID S. DIGERNESS PHOTO

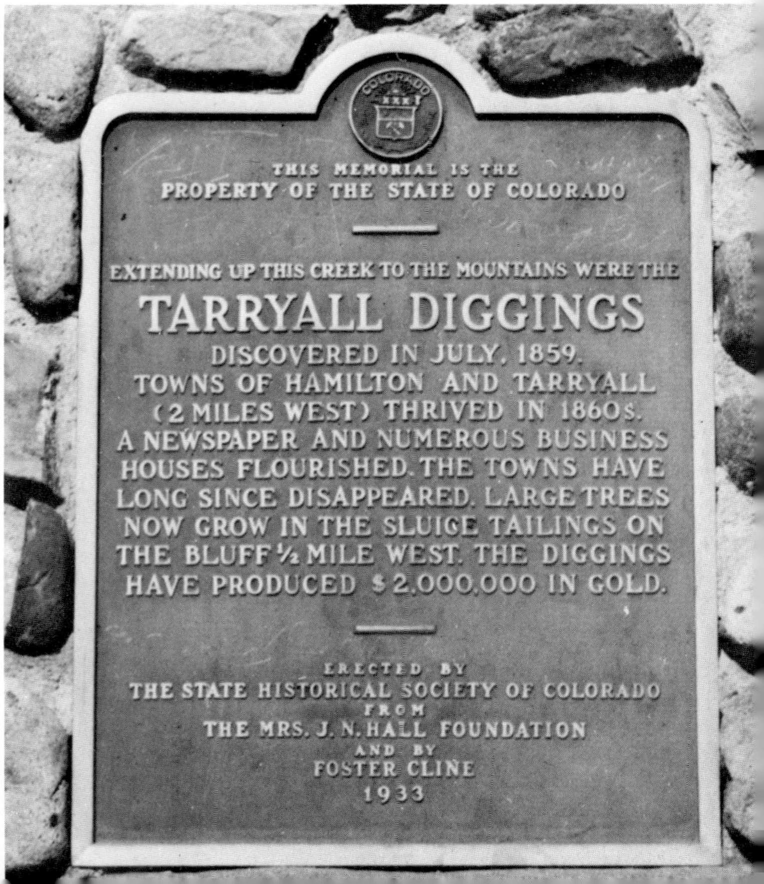

DAVID S. DIGERNESS PHOTO

DAVID S. DIGERNESS PHOTO

DAVID S. DIGERNESS COLLECTION

DREDGING OPERATIONS continued until recent years in old channels of the South Platte River, south of the town of Fairplay. This dredge boat was extracting gold from placer sediment deposits.

mines and the district. They, in turn, bought Rische's entire interest for $262,500, thus becoming half owners with Tabor in these claims and the Dives. Then, on November 18, 1878, the owners organized the Little Pittsburg Consolidated Mining Company. This included the New Discovery, Little Pittsburg, Dives and Winnemuck claims, the owners of the latter coming into the company on a basis of 27/100ths of the entire property and profits. Joseph C. Wilson was placed in charge of the consolidated mines company as general manager and superintendent. This made it the first large scale enterprise in the Leadville Mining District. Work was inaugurated on a much larger and more systematic scale than previously possible. Development of the consolidated mines started with the sinking of several shafts on the New Discovery and the Little Pittsburg. Shaft-houses and hoisting whims were constructed and an extra-large mine building was erected on the Little Pittsburg claim. After this, the property's production was greater than had ever been known in precious mineral mines, except for the Comstock lode in Nevada.

This great production was largely due to the enormous bonanza of mineral wealth penetrated by the shafts and drifts of the New Discovery. This claim, which had previously been somewhat behind, came out into the front and took the lead. The new underground workings penetrated an ore deposit of vast extent. This deposit — including both high-grade carbonate and low-grade iron ore — had a thickness of from 6 to 25 feet at nearly all levels explored. A few months prior to Jesse's visits, this spectacular bonanza showed a width in places of 30 feet. The uneven limestone foot wall had not been reached in much of the ground and there the ore extended still deeper. In a long level, a deposit of gray carbonates was opened that ran $100 per ton, which averaged 15 feet in thickness. The Little Pittsburg showed from 7 to more than 15 feet of ore in nearly all parts of the explored ground. There was much the same showing in the adjoining Winnemuck, and there was a considerable amount of unexplored ground. With such resources to draw upon, it did not take long to clear the cost of the Rische purchase.

Earlier company developments had proven that this was not a limited body of ore, but an enormous bonanza worth millions of dollars. The same great deposit was found to extend into the Little Chief, Carboniferous, Chrysolite and Vulture claims, and even far eastward into the Duncan, Robert E. Lee and others. What is remarkable is that the New Discovery production and much of that of other claims came

---

OPPOSITE, ABOVE and BELOW: The town of Fairplay appears to sparkle in the late afternoon sun. On the right side of this scene are tailings piles from the floating gold-dredge operation once common to the area. The DSP&P branch line from Garos served not only Fairplay, but also reached the mineralized area around Alma, northwest of Fairplay. A large gold dredge floats on a pond in South Park, near the town of Fairplay. This floating machine created the pond in which it worked—by scooping up great mouthfuls of gravel, processing it through the inner workings and dumping it onto tailings piles, such as the one seen at the right.

GURNSEY PHOTO
RICHARD RONZIO COLLECTION

"PLATTE STATION," or "Weston," was an early loading site for silver ore on the DSP&P. Ten tons of silver ore were stacked against the building and 65 tons of lead-silver bullion were piled on the bare ground, awaiting trans-shipment by rail.

PLACER MINING was being done by Chinese laborers near Tarryall in South Park when this photograph was recorded. These men were working parts of the abandoned diggings of previous prospectors with sluice boxes. The original owners of these "poor man's diggings" were in a hurry to get rich and leave the country. With a little time and patience, these miners could have joined together and worked the claims more successfully with flumes or hydraulics.

DAVID S. DIGERNESS COLLECTION

about almost entirely from development work, driving levels, etc. When the great ore deposits were being opened the previous fall and winter, it became evident that using horses for power was slow for even these short shafts of only from 40 to 80 feet. Four stationary steam engines were freighted to Leadville and the company had them in operation early in the spring of 1879.

Levels (tunnels leading from shafts) were driven forward in the mines so as to leave great blocks of mineral standing. It is interesting to note that the enormous size of the ore deposits called for a different style of shoring-up from that in general use. The so-called "Comstock square-set system" of timbering was adopted for supporting the walls and roofs of underground excavations. These sets consisted of upright hewn logs or sawed timbers, from a foot and one half to two feet thick, and from seven to eight feet high, fastened securely by similar, but shorter, cross-timbers above. In the thicker portions of these great ore deposits, three or four sets of these timbers were required, one above the other.

During April of 1879 alone, nearly $2-million worth of silver and lead was obtained from ore that brought the owners over $1-million in profit. It was figured that only two and one-tenth percent of the entire area of over 20 acres had been worked out or exhausted of ore. There was no evidence of the bonanza giving out near the producing shafts.

In the spring of 1879, the Little Pittsburg Consolidated Mining Company was incorporated in New York, with a capital stock of $20-million in $200 and $100 shares, and a large amount in $25 shares. The officers of the company included Jerome B. Chaffee, president; David H. Moffat, vice-president; George C. Lyman, secretary; and Charles C. Dodge, J. D. Smith, A. J. Dam, J. T. Scutter, C. L. Perkins, Henry Havemeyer, D. S. Draper, I. C. Babcock of New York, Senator W. H. Barnum of Connecticut, H. A. W. Tabor of both Leadville and Denver, James H. Chase, Chaffee and Moffat, as trustees. J. C. Wilson was the general manager of the mines.

Of those who originally were connected with these mines, Hook took his money and went back to Pennsylvania. Rische continued to live in Leadville, as well as in Denver part of the time, investing in mines, houses and business ventures. Before the Little Pittsburg was discovered, H. A. W. Tabor was postmaster and mayor of Leadville. In the early fall of 1878 — after the mine began to pay extensively — he sold his store and soon found himself rolling in money. He was nominated and elected Lieutenant Governor of Colorado, and subsequently lived part of the time in Denver, as well as in Leadville.

Leadville to Georgetown by Stagecoach

Jesse was so impressed with all he saw that he vowed he would return to Leadville as soon as possible. But he had to get back home to his family and his weekly newspaper in Georgetown. He took the stagecoach to Kokomo in the Ten-Mile country via Arkansas (Fremont) Pass.

Jesse got the privilege of riding alongside the driver of the stage and descriptively recorded this trip in his journal. He was in a position to really see the area along the wagon road. He described the area of the Ten-Mile District as one that consisted of converging slopes of two parallel mountain ranges with an intervening valley, through which Ten-Mile Creek flowed. The upper portion of this valley is almost a mile wide

DAVID S. DIGERNESS COLLECTION

and close to timberline. The westerly range contained most of the new mines, and is from 1,000 to 1,500 feet above the valley; it was named the Gore Range after "Lord Gore" — Sir St. George Gore, the Irish explorer — and farther north it is much higher and more spectacular, and is divided by the Colorado River. On the east, the Ten-Mile Range has many peaks from 13,000 to over 14,000 feet above sea level. The creek was called Ten-Mile because it was supposed to be 10 miles long, but in reality it is 17 miles in length. The two mountain ranges bordering Ten-Mile Valley extend northward from the main divide on either side of the saddle called Fremont Pass. The thrilling ride behind a six-horse team up the Arkansas Valley to the Pass was a distance of 14 miles north of Leadville; and, from the pass, water flows to different oceans. About two miles farther west along the divide, the Eagle River has its headwaters, starting from Tennessee Pass.

Just north of the pass, McNulty Gulch drains into Ten-Mile Creek near its source and the site of the new town of Carbonateville was one of the stops. The main gold production of this mining camp was in the early 1860's, with one mine still being worked by Colonel James McNassar, turning out an average of around $5,500 in gold each summer. It had a

AN OLD 10-STAMP MILL, which once was used for crushing ore, is shown here at the mouth of Trout Creek. In the background, Mount Princeton rises to an elevation of 14,197 feet, while Mount Columbia tops out at 14,073 feet — two of Colorado's highest peaks. Note the size of this stamp mill in comparison with the man standing to the left of it.

OPPOSITE: One of The Mineral Belt's pioneer wooden structures is the Sheldon Jackson Memorial Chapel at Fairplay. Dedicated in 1874 by the famed missionary, lecturer and explorer—who had established the church two years before—this little chapel is regarded by many as one of Colorado's most handsome early-day religious buildings.

DAVID S. DIGERNESS PHOTO

TOP
The
SHELDON JACKSON
MEMORIAL CHAPEL
A UNIT OF
South Park Community
(Presbyterian) Church

M. C. POOR PHOTO
DAVID S. DIGERNESS COLLECTION

GRADE CHANGES occurred along Trout Creek as the river bed was filled with sand. The first DSP&P railroad grade is nearly 15 feet beneath the sand at this point.

known yield from 1860 to 1879 of nearly $350,000. Farther down the Ten-Mile were the Follett placer diggings. The mining camp of Ten-Mile (or Summit, as it was also known) was one of the regular stops on the stagecoach run. Then, continuing further, the stage stopped at Kokomo, a mining camp that presented a strange jumble of log cabins, tents and primitive habitations. The activity of Kokomo intrigued Jesse and he vowed to spend some time there on a future trip through the Ten-Mile District. It was necessary to transfer to a different coach at this stop. Two stagecoach lines served the area, one from the south, the other from the north, and no direct through service was available at that time between Leadville and Georgetown. Today, the townsite of Kokomo is being buried under one of the huge tailings dumps of the Climax Molybdenum mine.

Jesse rode inside the stagecoach for the remainder of the trip down along the Ten-Mile to the point where it drains into the Blue River. From there the coach followed the Blue to the confluence of the Snake River. Thence, it headed upgrade along the Snake, making a stop at Montezuma. It then back-tracked a short distance to where Peru Creek joins the Snake and followed up along this stream through the mining camp of Decatur and crossed over Argentine Pass, the highest wagon road in the Mineral Belt. Jesse left Leadville at 5:00 a.m. and got into Georgetown at 7:00 p.m., using two different connecting stage lines for the trip, which cost $16. This included an added seasonal charge for clearing snow from roads over the high passes, a job done by men with shovels.

Leadville Express Train and Stage to Leadville

During the next few months, Jesse made several more trips to gather first-hand progress reports. On one of those trips, he reported further on the DSP&P. Riding on the newly established *Leadville Express* train, which left Denver late in the evening for an all-night trip, he recorded in his journal that the landscape could be seen under a full moon, which lighted the mountainsides like day at times. He enjoyed the relaxation of the trip, remarking that the coaches were dimly lighted with kerosene lamps. He knew some of the people on the train and they introduced him to a couple of men from Alma, who asked him to accompany them for a tour of their mine. There, they said he could get some new mineral specimens for his collection.

He noted that the railroad had completed the portion of the line over which he had earlier ridden on the work train and had since opened it to service. On May 19, the first passenger train had reached the top of Kenosha Summit. It was just getting daylight when Jesse's train arrived at the top of Kenosha Pass and stopped at a frame depot at Milepost 76.2. The locomotive had worked especially hard climbing up out of the valley of the North Fork of the South Platte River. A beautiful panorama unfolded as the morning sun burst above the horizon and illuminated the vast expanse of South Park. As the *Leadville Express* descended into this large upland mountain park, the glittering beauty of the snow-capped peaks ringing the valley came into view. Extending along the western side of the park are some of the grandest peaks of the Rocky Mountains: Mounts Guyot, Hamilton, Silver Heels, Lincoln, Buckskin, Horseshoe and others which range in elevation from 13,500 to over 14,000 feet.

Jesse recorded that about 10 miles more of railroad had been completed, with Jefferson as the terminus at Milepost 81, thus providing 81 miles of railroading and 60 of staging. The Mosquito Pass wagon road was completed on the first day of June, 1879. The stage companies had since transferred their lines to the new route. This cut nearly 20 miles of staging and reduced the fare. Roadbed construction in South Park was going at a fast speed and was expected to reach Fairplay late in the month. The town of Jefferson was the newest station on the railroad — with the elevation listed at 9,730 feet by the railroad survey and at 9,519 feet by Hayden's Government Survey. Jesse took the stage from Jefferson, going via Hamilton — at an elevation of 9,743 feet — to Fairplay, the county seat of Park County — 9,964 feet above sea level. There he got a room in the same hotel he had stayed in earlier. Lodging for one night and three meals cost $3.00. Jesse spent some time looking over the town and talking with a number of the town's prominent people, and had time to look over Thayer's placer operations along the South Platte River. Nearby was the largest such operation in Park County. The Fairplay placer property consisted of five miles of ground along the river, with dams and miles of ditches and flumes. On Thayer's leased ground, and under his management, nearly 100 Chinese immigrants were employed in hydraulic mining under their head man, Lin Sou. Jesse was told that a very rich streak in an old channel of the river had been

DAVID S. DIGERNESS COLLECTION

A DSP&P EASTBOUND PASSENGER TRAIN was working upgrade in Box Canyon heading for Trout Creek Pass in this photo of 1881.

M. E. CHASE PHOTO
STATE HISTORICAL SOCIETY OF COLORADO

TROUT CREEK flowed through an extremely narrow channel near its mouth, at "Devil's Cut," better known as Box Canyon. The DSP&P grade crossed back and forth across the creek bed in this arid area. This view near the Arkansas Valley was taken in 1888.

DAVID S. DIGERNESS PHOTO

MOUNT PRINCETON is one of the scenic wonders of the upper Arkansas Valley. Several miles behind the photographer's viewpoint the DSP&P crossed under Colorado's first standard-gauge mountain line, the Colorado Midland. Here, one sees the Arkansas Valley from the west side of Trout Creek Pass — not far from Buena Vista.

discovered a short time before his visit. He then returned to the hotel and had dinner.

Arrangements had earlier been made for Jesse to meet George Brunk and Assyria Hall, owners of the Dolly Varden mine, at the hotel. They headed for Alma and then went on to the mine where Jesse toured and observed operations and acquired some specimens. From the time the mine on Mount Bross was opened in 1872, it had paid a profit. It was known for the large amount of ore which returned from $300 to $700 per ton. The specimens Jesse got were of first-class ore which averaged 624 ounces of silver per ton. More ore had been turned out that year than ever before, the yield running to five and six tons a day. This averaged 257 ounces of silver per ton during one month, and nearly 300 another, for all classes combined. Twenty men were employed and from $3,000 to $8,000 had been cleared monthly. The Dolly Varden mine ranked second to the Moose mine in production and size of ore bodies, but often surpassed the Moose in richness of ore.

The Moose had been the most productive mine in Park County. It was first opened in 1871 and yielded $21,768 in the first 40 days. In the summer of 1872, mining was resumed, after being suspended during the winter, and was in full year-'round operation at the time of Jesse's visit. The Moose Mining Company consisted of 100 acres of surface ground, including 21 claims, as well as roasting and amalgamating works, which had two furnaces at the mining camp of Dudleyville. The Moose, like other mineral deposits of this section, was a series of deposits, some of them of great size.

EFDLEN PHOTO
RICHARD RONZIO COLLECTION

A SMALL ENTERPRISE called the Cornucopia Silver Mining Company ran this smelter at Buena Vista, circa 1886. The man on the left worked the scale, recording deliveries.

These were opened by tunnels and drifts. The total yield of silver for less than eight years, up to 1879, exceeded $1,150,000. The Moose was the first silver mine that attracted attention in 1871. The fall of that year saw a swarm of prospectors on Mount Bross and Mount Lincoln, and during the following summer, this was a great center of excitement. The mining towns of Alma and Dudleyville grew from the developments of these and other silver mines, and smelting works were soon in operation at both places.

The Mosquito Range, of course, includes Mount Bross and Mount Lincoln. The argentiferous riches of the Mosquito Range are found in limestone and sometimes in porphyry formations — a different condition from that existing among the northern mining districts of The Mineral Belt. Instead of true fissure veins, mineral chambers and pockets are often found between two formations. These deposits carry galena, gray copper, zinc blends and a quartz gangue, when located in the higher sandstone, and galena, gray copper and silver glance, and usually a heavy spar gangue, in the higher limestone of Mount Lincoln. Way above timberline, from 12,000 to 14,000 feet in elevation, mining was successfully conducted in the area.

The active ore-treating works in Park County comprised the Moose roasting and amalgamating works at Dudleyville, a stamp mill at Montgomery and dressing works at Alma. There were sampling mills, such as the Kendrick's, for the Golden smelting works and others. The Boston & Colorado Smelting Company had run a smelter earlier, but no longer smelted ores at Alma at the time of Jesse's visit; they only bought, sampled, and shipped to Argo north of Denver. Grose's wet concentrating works, built in Alma in 1876, dressed out 10 tons of ore daily. A quartz mill in the same building ran on ores from a number of mines, such as the Satisfaction, Phillips and others. The Phillips was the great gold lode mine in the southern portion of The Mineral Belt located in Buckskin Gulch. It was discovered in 1862; and, in a few short years, it produced around $300,000. Stamp mills were running full blast. The lively town of Buckskin Joe was in existence when the Phillips and Orphan Boy mines were yielding their thousands weekly. Some of the surface ore compared with the best in Gilpin County. When iron and copper pyrites were reached, work stopped because the millmen of those days were not successful in handling them. Some work had been done prior to Jesse's visit, and preparations were being made to operate the property on a large scale. About 1,200 feet of the lode had been opened by shafts, open cuts and levels, in development work which exposed the immense width between walls of the lode that ranged from 25 to 40 feet.

Horace Tabor told Jesse that he had helped to get Buckskin Joe started and that it was the county seat of Park County when he and his family were there from 1861 to 1866. He was the postmaster, and for awhile, the mayor. It was from here that they returned back over the Mosquito Range, to a renewed boom in California Gulch, at Oro City. Nearly all the inhabitants left Buckskin Joe, so the Tabors left, too. Tabor mentioned that he hated to leave for they (Tabor and about a thousand other people) had built a fine town and excellent courthouse. It was the Phillips lode, known as "the Diggings" that kept the town alive while it was booming. The town was loaded with saloons, as well as having a newspaper, many respectable businesses, a number of dancehalls and a theater. Old Man Dodge for a time got the town named Laurette for his wife and daughter, whose names were Laura and Jeanette, but that lasted only a short time and the more popular name of Buckskin Joe came back into use. The Orphan Boy mine, about a mile to the southwest, was owned by J. W. Smith and was a valuable lode since it was said to be on the same vein as the Phillips. Smith made a fortune in the mine in 1862 and 1863, but it was abandoned like the Phillips.

Mount Lincoln — named in honor of President Lincoln — lies north of Mount Bross, just south of the Mosquito Range's point of union with the Continental Divide, west of Hoosier and Breckinridge passes, the latter pass later known as Breckenridge, and still later, renamed Boreas Pass. The summit of Mount Lincoln is 14,284 feet above sea level and is one of the highest peaks in The Mineral Belt. The mountain is so situated, relative to the various ranges, parks and valleys, that the view is probably as fine as one will find in the area. There were a large number of mines on Mount Lincoln, some of which were rich and paid extremely well. The ores were of lower average grade than on Mount Bross, but the Ford and Musk Ox claims turned out very rich mineral. The ores carried galena, gray copper, sulpherets, brittle-silver and copper pyrites. The Russia produced the most silver, and between the time it was opened early in 1876 and the time of Jesse's visit, it had yielded close to $50,000 a year. The Danville mines were on the north spur of Mount Lincoln. The Present Help was at an elevation of 14,200 feet and was the highest mine in The Mineral Belt, as well as in North America.

At the head of the valley — to the east of Mount Lincoln in the Montgomery Mining District — is the old townsite of Montgomery, and located nearby were the Chicago, Wheeler and Atlantic mines, the last two having yielded a great deal of free-milling gold quartz.

Besides getting a good number of mineral specimens from the Dolly Varden, Jesse managed to get some other fine specimens from other mines of the Alma area. It was dark when he arrived back in Fairplay. After eating, he specially packed each specimen into individual containers and placed them into a trunk that he had brought along just for that purpose. The stage line charged 10 cents per pound for all baggage in excess of 40 pounds to the passenger.

Over Mosquito Pass by Stagecoach

The following day, he was on one of two Concord coaches, each drawn by six-horse teams, which rolled out of Fairplay at the same time, headed for Leadville. Unfortunately, Jesse was aboard the stage following the first one and found — along with the other passengers — that he was in a swirling mass of dust with no way to escape. The coach he was aboard stopped in Alma long enough so that the first coach had a chance to move out ahead by at least 10 minutes. Thereafter the dust ceased to be a problem as the coach rolled along upgrade from Alma, climbing up to the town of Mosquito, where a combination hotel, eating and stage stop served all the stage lines. Located in the Mosquito Mining District of Park County, this settlement dates back to the early days when it was located on a trail over the Mosquito Range between the towns of Hamilton and Oro City. The Forest Queen mine, located in this mining district, had ore which milled from 158 ounces of silver per ton to 690 ounces per ton. The Champagne mine was another valuable lode, as was the Joe Chaffee, carrying silver and gold. Production was good in both. The London mine was coming ahead in this district at the time of Jesse's visit. The Kansas was another one of the leading mines and showed large deposits of rich ore. As the transportation facilities in the surrounding mountain area

POINTING HIS CAMERA toward the west—over the town of Buena Vista—L. C. McClure recorded the view seen by Colorado Midland passengers in 1898. On this page, the depot of the D&RG can be seen behind the

L. C. McCLURE PHOTO
DAVID S. DIGERNESS COLLECTION

business buildings on the lefthand side of the main street, while the C&S depot is plainly visible on the outskirts of town—in front of the courthouse. The buildings in front of the C&S depot belong to the Kunsel smelter. In the background, the majestic bulk of Mount Columbia rises to an elevation of 14,073 feet—part of the Great Divide. Leadville—the goal of both the South Park and Rio Grande railroads—is to the north (right).

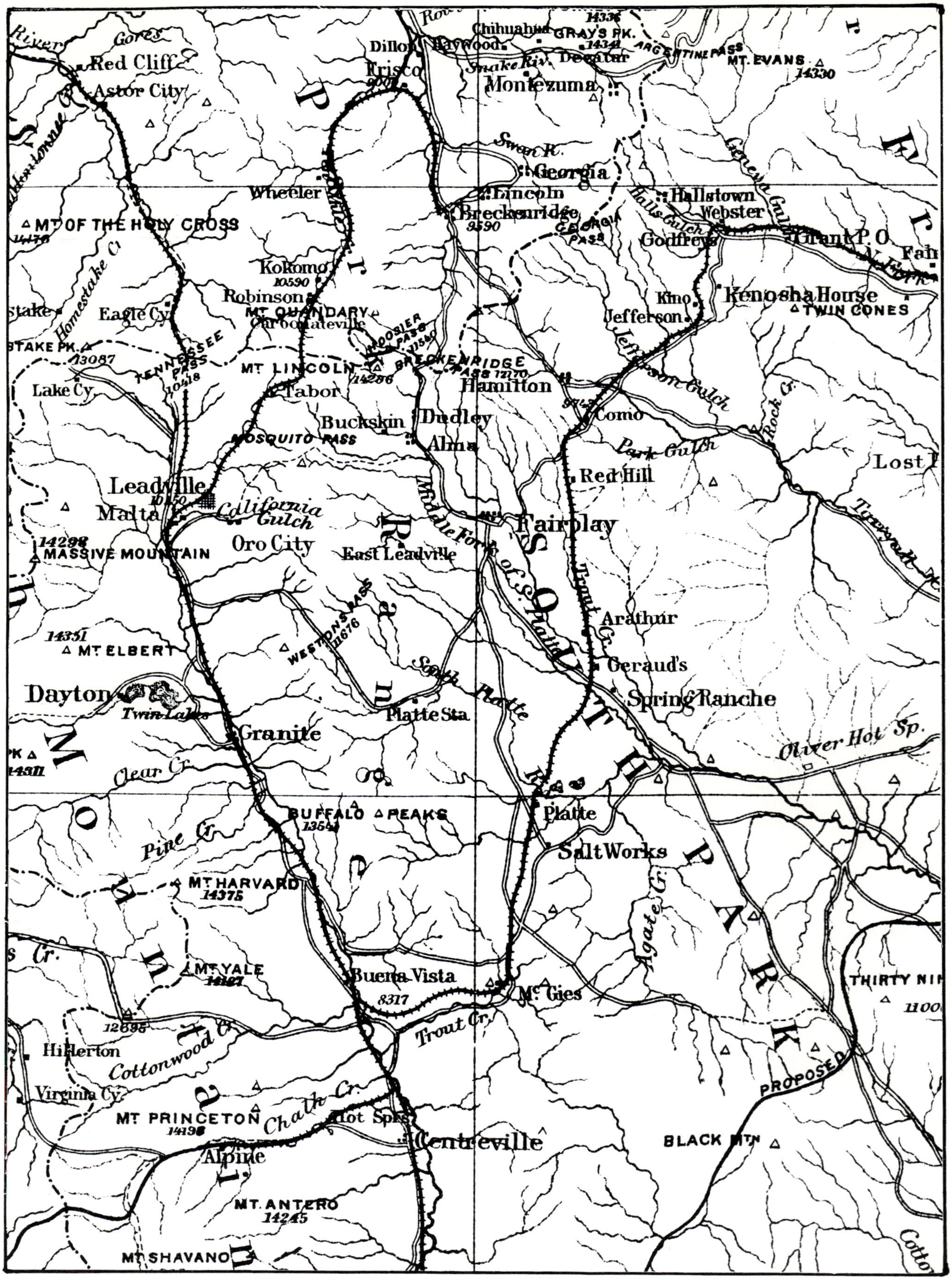

MAP OF THE OLD SOUTH PARK RAILROAD, WEBSTER TO BUENA VISTA.

STATE HISTORICAL SOCIETY OF COLORADO

AN AMBITIOUS PHOTOGRAPHER gathered a group of 40 men together in 1885 to pose them atop and around Mason bogie Number 40, which was resting on the DSP&P's turntable in Buena Vista. Number 40 originally was built in May of 1878 for the DSP&P as their Number 3, and was named the "Oro City."

improved, many miners from the mining camps around Alma moved into that town, even though other camps were there first. Alma was growing and seemed to be located in the best position to serve all the surrounding area.

The Mosquito Pass wagon road had just been opened from Alma to Leadville early in June. It brought considerable travel to Alma and also to the old town of Mosquito, located on the east side of the pass. The pass was named for the town of Mosquito, as was the Park Range of mountains, which just recently had been renamed the Mosquito Range. Jesse found that the large placer claim on the South Platte River at Alma was that of Mills, Hodges & Company. It yielded around $65,000 in gold altogether since this company took control of it, and was paying better than ever at the time of Jesse's tour.

The stagecoach stopped at Mosquito, where the first coach also had stopped. The eating house was large enough to handle all the business. The other coach pulled out at least 15 minutes ahead of the one that Jesse was on. Back on the stagecoach, Jesse was allowed to ride on top beside the driver, who proved to be a talkative man, gentle and courteous, as well as very careful and kind to his horses. The alpine countryside reminded Jesse of Argentine Pass, which ranked as the highest wagon road in The Mineral Belt, followed closely by this pass. The rattling coach cheerfully rolled along — first on one wheel and then on another — as it glided behind the six horses that dashed wildly along, climbing the steep, rocky pass, until it surmounted the range of mountains at Mosquito Pass. Down the other side, the coach ran along the precipitous mountain road. It provided a beautiful view of Mount Elbert and Mount Massive, across the wide Arkansas Valley — a sight seen all the way down the pass.

It was marvelous how the horses helped hold back the coach on the steep downhill grade. The brakes on the wheels of the coach were used almost continuously as well. The friction of the wheels on the rocks caused sparks to fly. At times the coach would run very rapidly coming down. It really was a spectacular ride; the coach spun down the steep winding road, surging and leaping back and forth. The driver mentioned that "if you do not understand horses at all, especially 'weak minded' fellows, they wouldn't allow you to hold that brake. A few days ago, in going down a hill like this, I pressed the brake with my foot, they got mad and snapped their check reins, seized that neck-yoke with their teeth, and fairly set the ground afire, pitching down that hill. You must treat a horse according to his education or you will always have a rebellion on your hands." A mention of the number of wagons passed might be of interest. These were all tense moments that made the passengers' hair stand on end. The new stage route was heavily patronized and now was one of the main thoroughfares to Leadville. Before the wagon road was opened, only a trail existed, which accommodated single-file animals or foot traffic. This pass is where a number of travelers froze to death in the winter during the silver rush to Leadville. No one could survive the terrific snowstorms which sweep the mountain passes, and if an unsuspecting traveler were caught in one, he would surely perish. The stagecoach rumbled through Evans Gulch and shortly was in Leadville. As the passengers climbed down from the coach, they seemed to feel a kind of relief that the trip was over. After trying other hotels without success, Jesse finally rented a room in the Clarendon Hotel, located on Harrison Avenue, and then set off to visit the Little Pittsburg Consolidated Mining Company, Inc. The following tabulated statements were supplied from the company books

CHARLES WEITFLE PHOTO
MR. AND MRS. ERNEST GERBAZ COLLECTION

OPPOSITE: The Box Canyon of Trout Creek was the setting for this dramatic W. H. Jackson photograph of 1887. Jackson was heading westward, en route to Nathrop with his special train. This rocky canyon was a treacherous place during heavy rainstorms; flash floods could sweep tons of sand and gravel down the Trout Creek drainage, covering the track with debris.

W. H. JACKSON PHOTO
DAVID S. DIGERNESS COLLECTION

THE NEW IRON-TRUSS BRIDGE near Milepost 133 — just east of Buena Vista — was the setting for this South Park train. The main span of this bridge over the Arkansas River was 152 feet long, with two wooden-bridge structures connecting each end, making the structure 293 feet in length. The track was 48 feet above the river. Mason-bogie locomotive 22, a 2-6-6T named the "Crested Butte," was nearly new in this 1880 view.

DENVER PUBLIC LIBRARY, WESTERN HISTORY DEPARTMENT

A DSP&P LOCOMOTIVE was switching the yards at Buena Vista in this view of 1881. The DSP&P owned one-third of the Buena Vista Townsite Company, and, in conjunction with them, built this hotel.

and show production from discovery to November 18, 1878, amounting to $900,000 from 7,000 tons of ore, which yielded $588,000 to the owners. Production figures included the Little Pittsburg, New Discovery and the Winnemuck mines, some of the most valuable properties in the district.

From November 18, 1878, to April 18, 1879, a total of 7,807 tons of ore were sold, with receipts to owners of $484,984 and actual yield of ore sold or assayed at $1,078,586.

From April 18 until June 18, 1879, the Little Pittsburg Consolidated Mining Company sold a total of 7,900 tons of ore with receipts to owners of $480,748 and actual yield to smelters of $1,106,000, with a grand total of 25,007 tons and $1,553,732 in receipts to owners and actual yield to smelters of $3,084,586.

Several mining experts made a thorough examination of these mines. Professor R. W. Raymond estimated that the high-grade reserve ores alone would sell to the smelters for $2,000,000. The New Discovery, with an opening of 300 feet by 100 feet by levels and cross-cuts, was figured at 13,000 tons at $1,100,000 — nine cubic feet to a ton of ore. In ground explored by a long cross-cut, 4,000 tons of gray carbonate ore was figured at $400,000; the Little Pittsburg was tabbed $350,000, and the Winnemuck at $50,000. Estimated receipts for low-grade ore on the dumps totaled $100,000. This did not include the contents of any unexplored ground. The assay value of the ground opened approached $4-million.

An interesting development of great significance was reported in the New Discovery. In a new shaft sunk near the line of the Vulture and Chrysolite, they found what appeared to be a new and lower strata of ore body. The new strata was reported to be richer even than the one above. It was evident that the Fryer Hill mines possessed a double value, and it was thought that the whole area might be underlaid with two or more stratifications of ore.

For 15 years, Jerome B. Chaffee, president of the Incorporated Pittsburg Consolidated Mining Company, had been a leader and controlling spirit of the Republican party of Colorado. He was a legislator from Gilpin County in the earlier territorial days, and was speaker of the House of Representatives. In 1865-'66, after he had made a fortune on the Bobtail mine in Gilpin County, he was chosen United States Senator, but the state had not yet been admitted to the Union at that time. He was delegate to Congress from 1871 until 1875, and became one of the first two United States Senators from Colorado when the territory became a state in 1876. (The other U. S. Senator was Henry M. Teller of Central City.)

David H. Moffat, who was vice-president of the Incorporated Company, had been equally prominent in financial operations. Since 1865 he had been cashier of the First National Bank of Denver, of which J. B. Chaffee was president, and had been connected with the latter in many mining operations. He was treasurer of three railroads operating out of Denver and held important territorial and state offices.

At this time, J. C. Wilson resigned the office of Internal

OTTO C. PERRY PHOTO
DAVID S. DIGERNESS COLLECTION

COLORADO & SOUTHERN Number 62 was resting in Buena Vista when Otto Perry shot this picture. In all too short a time, this 2-8-0 would be no more.

Revenue Collector of Colorado, to take charge of these great mines. He was president of the convention that framed the state constitution.

Jesse met all these people; however, Horace Tabor was his best source of information.

When the Little Pittsburg harvest was rolling out great wealth, Tabor started to buy all the claims around him that could be picked up at reasonable prices and where probabilities existed regarding future production. There were no outcroppings to indicate what was beneath the surface, and it required considerable risk and confidence to pay all the way from $10,000 to $50,000 at a time for undeveloped and unknown ground. Tabor "lucked out" more often than not, and the subsequent profits he made from this ground proved to be tremendous. But few would have made those investments at the time he did. The claims, or mines, referred to are the Carboniferous, Chrysolite, Little Eva, and part of the Vulture, Colorado Chief and the Fairview. Mr. Field probably had as valuable an interest in these mines as his partner in the Chicago Dry Goods business had in the Iron mine. The profits from November of 1878 until Jesse's visit had been from $20,000 to $40,000 monthly.

The Carboniferous was the first developed of these properties. The underground workings of this mine and of the Little Chief opened into one another, and the same body of ore was found to be from 22 to 30 feet thick in one locality.

The Little Chief mine was bounded by the Little Pittsburg on the east, and by the New Discovery and Carboniferous on the west. It was located by four men: Patrick and Richard Dillon, Peter Finnerty and John Taylor. Soon afterward, Tabor, his partners and Fryer started to take out ore at an extremely fast rate. These men struck carbonates in their mine; and, in a few months, they had more money than they knew what to do with. They received about $100,000 for ore, and in December of 1878, sold to J. V. Farwell and associates of Chicago for $300,000. When the purchase was made, the underground workings were in poor condition and unsafe in one locality. The new management went right to work to place the mine in shape, and shortly thereafter, reopened it with a number of excellent shafts, supplied with double cages and cars. Specious buildings were erected and three stationary steam engines, hoisting machinery and a smelter were installed and put into immediate service.

The ore deposits of the Little Chief opened during the previous winter were of great proportions, rivaling some of the best ore on the hill. Long drifts were driven 10 feet wide and 22 feet high in ore that brought $100 per ton and some lode matter extended between ore on either side. The yield had been somewhat larger since the installation and completion of all the improvements. There was one area in the mine where the ore ran nearly 30 feet thick. Work was carried on with steam machinery through three shafts and the profits were extremely good. They had close to 90 men employed under the direction of F. J. Folsom. The lode carrying the valuable mineral was from 30 to 100 feet below the surface. One of the shafts had a depth of 115 feet, called Shaft Number 1; from there, the lode dipped downward to both the east and west, along porphyry rock serving as the base, the sides of which contained the ore deposit. Here there were black-sand carbonates averaging $150 a ton, and so soft that just a pick was needed to break it down.

Now let's get back to the Carboniferous mine. The same method of shoring-up was used as in adjacent claims. It was necessary to sink a shaft to a depth of 115 feet below the surface in order to reach the limestone formation. There were

---

OVERLEAF: In 1881, the Buena Vista transfer yards were shared by both the D&RG and the DSP&P. Two South Park locomotives—equipped with large Congdon smokestacks—were on the house track at the depot, waiting for a switch-engine to make up their train. From here, they would proceed toward Leadville, northward to the end-of-track at Malta. Silver ingots were piled high on the depot platform, awaiting shipment. The main street ran parallel to the yard tracks at this time. To the left, what appears to be great piles of coal or coke had been shoveled out of railway cars. It then had to be reloaded into freight wagons and hauled to the Leadville smelters.

BUENA VISTA LAND CO.

W. H. JACKSON PHOTO
RICHARD RONZIO COLLECTION

BUENA VISTA'S main street (to the left) was nearly new in this photograph, taken about 1883. Mount Yale rises above the town (to the left) and a shoulder of Mount Columbia is in view (to the far right). Two passenger trains were in town on this day — visible at the far end of the main street. Both D&RG and DSP&P locomotives can

W. H. JACKSON PHOTO
RICHARD RONZIO COLLECTION

be seen over the rooftops, adjacent to the Rio Grande depot. This scene was typical of the period when both rail lines shared the trackage to Leadville.

C. W. ERDLEN PHOTO
RICHARD RONZIO COLLECTION

THE MAIN STREET of Buena Vista was the setting for a parade welcoming some distinguished visitor(s) when this photograph was taken in the 1880's. An enterprising depot agent had remodeled the attic of the D&RG depot,

and was using it for living quarters. At the time C. W. Erdlen shot this view, the DSP&P and the D&RG shared trackage between the mining centers of Buena Vista and Leadville — via the upper Arkansas Valley.

W. G. CHAMBERLAIN PHOTO
RICHARD RONZIO COLLECTION

GRANITE CITY, on the Arkansas River, was a booming mining town and shipping point in 1876 — years before the arrival of the railway. Every day, more than 800 6-, 8- and 10-mule teams hauled wagons through Granite, on their way to the great silver bonanza of Leadville. At the time W. G. Chamberlain recorded this view, Granite was the county seat of Lake County. At a later date, the trail heading off to the west (left) gained importance as the main route to and from the Aspen Mining District.

OPPOSITE: Cottonwood Lake is located a short distance west of Buena Vista—on Cottonwood Creek in the San Isabel National Forest.

DAVID S. DIGERNESS PHOTO

RICHARD RONZIO COLLECTION

GRANITE, COLORADO, was an important ore-loading point for the Denver & Rio Grande narrow-gauge railway during the 1880's. With winter snow on the ground, freight haulers converted their wagons to sleds with ski runners. Sacked ore of untold value had been stacked on the freight dock. The Denver, South Park & Pacific also used this station, and shared the tracks with the D&RG between Nathrop and Leadville.

usually five feet of hard carbonate rock, and the remainder of the lode was a material called sand carbonate. In the first five months of work, 2,890 tons of ore were mined, hoisted and sold. The yield ranged from 35 to 570 ounces of silver per ton. Much of the ore sold at $100 per ton.

The Chrysolite mine adjoined the Carboniferous on the west, and had much the same quantity and quality of ore. The main shaft was over 100 feet deep. The yield of this claim in two or three months prior to January 1, 1879, was about $50,000. The ore carried over 20-percent lead. In the Vulture mine, shafts had been sunk, but no ore was found in it or adjacent claims until late in the fall of 1878. The prospecting had been done on the wrong slope of the hill. An experienced lead miner from Missouri believed from the lay of the ground that ore would be found on the northern slope. The first shaft in that locality — at a depth of 47 feet — revealed one of the best-paying ore deposits in the district. Tabor, Borden & Company soon bought a two-thirds interest in the Vulture. The ore was from 15 to 18 feet thick. Much of it yielded from 80 to 200 ounces of silver per ton, and was from 30- to 70-percent lead. For a time nine-tenths of the ore mined was rich enough to sell direct to the smelters.

At the time when many of these claims were surveyed, a small plot of ground had been overlooked and was not included in the surveys of adjoining claims. It was of a triangular shape adjoining the Vulture and Little Eva. It was 35 feet wide by 60 feet long on its side and was named the Triangle mine. Its placement allowed mining of the area between the Vulture and Little Eva, including only 2,200 square feet of surface ground. It was underlaid with the same Fryer Hill bonanza which produced ore that sold for $38,000 in less than three months. During the winter, a part interest sold for $16,000. Its total yield was reported at $58,000 and cost less than $6,000 to exhaust the ore. Its shaft was used for access in mining the profitable Little Eva claim. Not far away was another even-smaller unclaimed spot of ground, about 60 feet long and from one to six feet wide. This was named the Silver mine.

The Buckeye, O. K., Pandora, Muldoon and Hope Claims also were on Fryer Hill.

The Amie mine was parallel to and on the eastern side of the Little Pittsburg. The location originally was surveyed about the same time. Scattered seams of mineral were found at intervals all the way down three shafts, but they were not drifted on, and were not generally large enough to pay. The owners knew their claim was too near rich deposits to be generally barren; therefore, another shaft was sunk further north. This one struck the desired carbonate layer at 140 feet, which was rich enough and extensive enough to insure fortunes for the owners. Two or three weeks after the ore was penetrated, $20,000 was cleared from sales of ore taken out. The ore deposit was four feet thick and of very high value.

The Climax mine was on Fryer Hill, east of the Amie and west of the Duncan. It was owned by Berdell, Witherell and others, and was superintended by Paul S. Ross. This mine was opened by three shafts and had many levels. Shaft Number 1 was down 320 feet or more, and in the bottom was valuable ore. Shaft Number 2 was 190 feet deep and cut an 18-inch strata of carbonates 120 feet deep, then passed through layers of porphyry and limestone. The third shaft was 140 feet deep. Quite a large amount of ore had been mined and some of the better classes turned out from $800 to $1,000 per ton. Only portions of the workings showed lode matter of value.

The Duncan was on the eastern part of Fryer Hill; and, like some of its neighbors, was remarkable for the extent and value of its ore deposits. These were not encountered in large quantities at the start of development, but perseverance finally led the way to fortune.

Ore from the Duncan milled from 40 to 500 ounces of silver to the ton. The claim was 1,360 feet long by 300 feet wide, while the shaft needed to reach an ore deposit eight feet thick was 112 feet in depth.

The famous Matchless mine was located between three mines — the Duncan, the Robert E. Lee and the Union — and the same rich body of ore extended into all of these other claims. At a depth of 100 feet, the shaft entered a body of ore running 60 ounces per ton. Valuable sand carbonates were found in drifting southeast. A drift following the foot wall of silicate of lime with galena carried 60 ounces of silver, and finally entered into ore of better-than-average grade. The strata was similar to that of the Little Pittsburg. The Matchless was Horace Tabor's most-prized possession and was one of the great producers of the district. Tabor bought the property in 1881 for $117,000 and it yielded him some $10-million over the years, paying $100,000 a month at times. One shipment of ore from the Matchless assayed at 10,000 ounces of silver per ton.

The so-called Tabor cabin is just south of the Matchless shafthouse and is a tourist attraction of the region. The second Mrs. Tabor — the renowned *Baby Doe,* a celebrated beauty of the mining camp — lived alone in this cabin for many years after her husband's death. (The Baby Doe saga will be set forth in detail in Volume II of *The Mineral Belt.*)

At the time of Jesse's visit, the Robert E. Lee was the newest wonder of this district. A bonanza of unsurpassed richness, the Robert E. Lee had a fabulous lode, ranging from 10 to 12 feet thick, which yielded from $200 to $400 a ton — the highest grade ore in the district. This property was claimed by a party of Colorado Springs men, in addition to Henry and E. O. Wolcott and others. The former obtained possession of the mine and held it by force of arms. The latter got out an injunction to stop work, and this was the way matters stood when Jerome B. Chaffee and others stepped in and purchased the mine for $240,000, all claimants getting portions of the money. Trouble among the old owners was all that caused them to dispose of it at that figure, as it was asserted that the purchase money could be taken out of the mine in a month or two. This was a continuation of the same Little Pittsburg and Duncan ore deposit, but it was richer here than elsewhere. During one 3-month period, over $500,000 in silver ore was mined; in 24 hours alone, nearly $118,500 was taken out. Besides Jerome B. Chaffee, the purchasers of the Robert E. Lee were David H. Moffat, Senator J. P. Jones of Nevada, Congressman S. B. Elkins of New Mexico and the Delmonicos of New York.

During this visit, Jesse wanted to see the placer-mining operations that were an important industry in California Gulch. A few years prior to his tour of the area, water had been brought in from streams miles away by means of ditches and flumes, as mentioned previously. This water supply permitted the use of hydraulics, which washed down huge quantities of gravel and stream-bank so speedily and cheaply that heavy profits were obtainable from ground carrying but a small amount of gold. The most important enterprise, of course, was that of the Oro Ditch & Fluming Company, which was managed by W. H. Stevens. Several Little Giant hydraulics were used in this operation. The claims of Thomas Starr and of Stevens and Leiter were worked in the same manner. The growing wealth and capacity of the district was shown in the fact that more than a dozen new mines

EVERGREEN LAKE.

DAVID S. DIGERNESS COLLECTION

DELL A. McCOY PHOTO

THE GRAND SCENERY of the Rockies did not always make up for hardships along the road — as travelers headed westward above Granite, toward Aspen. During the early days of settlement, the greatest challenge was the crossing of Independence Pass, high above this scene. Lake Creek flows into Twin Lakes near this view.

DELL A. McCOY PHOTO

LAKE CREEK tumbles through a chasm as it flows toward Twin Lakes. This autumn scene was familiar to the many prospectors and freighters who traveled the road on their way to Aspen from Granite. The little settlement of Granite was the nearest railhead for several years, a busy place during that period.

THE TOWN OF GRANITE was well-settled in 1885, with both the DSP&P and the D&RG railroads serving the community. The main business section was across the Arkansas River (to the east), featuring a hotel, a blacksmith shop, restaurants and a depot. To the right of the D&RG depot are several freight sleds, while on the near

C. W. ERDLEN PHOTO
DAVID S. DIGERNESS COLLECTION

side of the river stand wagons, which were used during the summer season. Several narrow-gauge freight cars have been "spotted" on sidings beside the joint D&RG-DSP&P mainline.

developed into profitable enterprises during the spring months of 1879, while a number of others that had previously displayed but little mineral, produced on a regular basis for a considerable length of time. Some of these possessed ore deposits of tremendous size and value.

Jesse was after unusual specimens, and his quest took him to many of the Leadville mines and smelters. On Carbonate Hill many new mines had discovered great wealth; also, the area extending northward to near the head of Stray Horse Gulch contained what was known as the famous Iron mine zone. On the slope rising from California Gulch — nearly two miles above Leadville and a short distance from Oro — were the Lime, Bull's Eye, Smuggler and other mines. Not far from the center, and on the western slope of the hill, was the great Iron mine, and east of it were such properties as the Argentine and Adelaide. The Iron mine was considered by miners and mineralogical experts as a typical carbonate deposit, the most regular in form and continuous in extent of any in the district. There were (and probably still are) large deposits of ore elsewhere, but none more uniform. The Belle of Colorado was the next claim east of the Iron. It was owned by Joseph Pierce and H. A. W. Tabor.

The owners of the Iron mine were W. H. Stevens and L. Z. Leiter. They had a serious battle with adjoining claimants at the time of Jesse's visit. This mine was the first to be systematically opened and developed. From the mine shaft, levels led north and south at intervals of 100 feet. These levels were hundreds of feet in length and altogether aggregated several thousands of feet. Some of them followed the dip and direction of the deposits, and all showed ore of either high or low grade. One drift passed through 100 feet of carbonate ore of a gray complexion, largely composed of 300-ounce mineral. Another followed the pitch of the contact waves for over 400 feet, crossing other openings diagonally. Others were driven for shorter distances at right angles, making use of cross-cuts to open the ground, allowing great blocks of ore to be extracted. Above the main ore deposits was an area of chalky porphyry, jointed in crystalline forms, with no fragments of lime intermixed, as in many other claims, and below was the solid, unbroken limestone foot-wall, with a uniform general dip eastward in a series of undulations, slopes, and steps. In one place there was a solid mass of galena, and there were streaks and patches of chlorides and galena scattered in nearly all parts of the mine. The deeply-tinged iron often was of low grade, but there were hard carbonates in very large quantities, as well as those of softer character. The Iron claim comprised a surface location 1,500 feet by 300 feet wide, up and down the hill.

During the fall of 1878, a miner by the name of Williams went up the hill beyond the property line of the Iron mine and sank a shaft in order to strike the known deposit, which was found at a depth of 200 feet below the surface, and 75 feet west of the end of the Iron incline. He then barricaded the latter at that point, and by force of arms, held the interior workings against the Stevens and Leiter miners. During that winter, a force of old Georgetown miners were employed by Stevens to work for Williams as guards at his mine — named the Law Claim — and they obtained the barricaded workings for Stevens. The mine, in turn, was heavily guarded against Williams. In the meantime, the dispute came up before the courts, the Iron owners claiming they could follow their ore vein wherever it led, while the other side asserted that there was no vein in the true sense of the term, but only a deposit, which could not be followed beyond side lines. The final trial of this case was decided in favor of Stevens and Leiter of the Iron mine.

TWIN LAKES HYDRAULIC MINING.

DAVID S. DIGERNESS COLLECTION

OPPOSITE: Mount Yale, at 14,196 feet above sea level, is part of the Collegiate Range near Buena Vista, and is one of Colorado's famous "fourteeners." Here, the viewer is looking along Cottonwood Lake road, which runs through a portion of San Isabel National Forest.

DAVID S. DIGERNESS PHOTO

MR. AND MRS. ERNEST GERBAZ COLLECTION

"ON TO LEADVILLE," the legend for this 1879 view read. Stagecoaches hauled by six-horse teams — run by McClellan & Spotswood's South Park Stagecoach Line — met the DSP&P trains at the end-of-track. The railhead was being pushed toward Leadville when this view was taken north of Buena Vista.

Chapter VIII

The Leadville Mining District

A REPORT OF SILVER ORES smeltered or shipped from the Alpine Mining District interested Jesse and he decided he would like to obtain some specimens from this district. The Alpine Mining District was in a region that had been part of the southern portion of Lake County, in an area that was set off for a new county named Chaffee, which also included Lake County's old county seat, the town of Granite. Chaffee County comprises a large portion of the upper Arkansas River Valley and is south of the present Lake County. The town of Alpine, from which the Alpine Mining District got its name, was 60 miles south of Leadville. Some of the best silver lodes were located in this area, along Chalk Creek. Jesse got some small specimens from the smelter, which came from ore mined in the Alpine Silver Mining Company. Some of the other specimens he secured were mine crystals that formed in air pockets in the veins. These included galena, various oxides of iron and manganese, and small amounts of chloride of silver.

The Leadville Mining District defied all previously developed rules regarding mineral in place, or in deposit. In this district, mineral is found by sinking shafts, not by outcroppings, as in mineral veins elsewhere. The objective point in the contact of the porphyry and limestone — between which formations the mineral is almost universally encountered — is the iron, which is a sort of cap rock above the ore, and is called the "contact." Below this, if no ore is found, and otherwise beneath the ore, blue limestone usually occurs in undulations like a disturbed sea. If no ore is visible at such locations, the miner drifts on the contact with the hope of finding less barren localities beyond. Rich ore deposits have been found by this method. In addition, some carbonate mines yield gold in small amounts.

The Grant smelter was erected by J. B. Grant, and was situated in the lower portion of Leadville. The first furnace was fired up October 1, 1878, and continued without interruption until the time of Jesse's visit. On the first of January, a second furnace was fired up. So great were the demands upon the smelter that two more furnaces were added, one after another in the spring. Unlike many similar establishments, no furnace troubles or interruptions of any kind were encountered. The four furnaces handled from 85 to 95 tons of ore daily and turned out from 20 to 30 tons of lead or base bullion. This of course contained rich silver and was shipped in the form of bars to be refined later, when the silver would be separated from the lead.

J. B. Grant superintended the works from the beginning. Here, shortly after Jesse's tour, they handled the entire ore product of the Little Pittsburg Company, which necessitated the addition of four more furnaces. The ores of this district were treated by smelting without roasting, nature having obviated the necessity of the latter. The ores were smelted in what were called water-jacket furnaces. These were constructed of iron, and were circular in shape, about six feet in diameter, lined internally with fire clay, and rested on a cement-and-clay foundation.

Charges of ore and coke or charcoal were shoveled in from a floor above the point where the bullion was discharged. The furnaces were upright, and extended upward through the building, with outlets for fumes and smoke above. The proper mixture of ores and the requisite proportions of fuel were important factors in the success of this type of operation; therefore, the more refractory the ores, the more necessary was careful operation of the furnace to avoid disaster — such as chilling the furnace and other troubles. Weighing the ores was one means of determining their character. Into these furnaces were fed charges, consisting (as mentioned) of the proper proportions of ore, coke, charcoal and slag — slag being the refuse from previous charges. The molten mass separated itself according to its specific gravity; lead — with its silver content — settled in the lead-well at one side and near the bottom of the furnace. From this well, the bullion was dipped into iron moulds where it cooled into bars that each weighed about 100 pounds. Each furnace turned out from 60 to 100 of these bars, or from 6,000 to 10,000 pounds daily, $400 worth of silver carried in each ton of bullion. The water-jacket furnace was prevented from succumbing to the terrific heat by the continual circulation of water between the inner and outer plates of its construction.

The directors and sub-directors of such works, as well as the smelter workers, held very responsible places, for ignorance, inattention or lack of skill was likely to throw everything out of gear, chill the furnace and stop operations for an indefinite period. The feeder was obliged to exercise equal care and judgment, and was subject to that great evil — being poisoned

PANORAM

DAVID S. DIGERNESS COLLECTION

by lead fumes. A melter and other employees were required below the place where the lead bullion was taken from one outlet and the refuse, or iron slag, from another.

A furnace such as this was run night and day from one month's end to another; to allow it to cool down would have entailed a heavy expense in drilling out the mass of iron and slag that would have to be removed and, in fact, would stop business completely. The fuels used were charcoal and coke, the former bought from the adjoining forests at 16 cents a bushel, and the latter shipped by rail, then wagon train over Ute Pass from Trinidad and El Moro, over 200 miles away, which cost $28 to $35 a ton. The price was $68 for a time during the freight blockade in the spring of 1878.

To smelt each ton of ore, something like 40 bushels, or 560 pounds, of charcoal, worth $6.40 in 1878, were required, along with about 660 pounds of coke, worth $10 or more. In blockade times, coke cost from $60 to $70 per ton. These works contained a 50 horsepower steam engine, two tubular boilers and the usual crushing-and-drying apparatus.

Another smelter where Jesse got some mineral specimens was the Harrison Reduction Works, the first in operation in Leadville. Jesse observed that there were over 1,500 tons of ore on hand ready for treatment at this smelter. He also obtained specimens at the Malta, the Adelaide, the Berdell & Witherell La Plata Company, Germania, Iowa, Garbutt, Saint Joseph and the Little Chief smelters. Also, the sampling and ore-buying works of A. R. Meyers & Company and Eddy & James had some fine mineral specimens that Jesse acquired.

In talking with Tabor, Jesse — being interested in details — wanted to know what his middle names were. Austin Warren was the reply. So Jesse wrote out Tabor's full name: Horace Austin Warren Tabor. Tabor said, "Just call me 'Haw' and call my wife 'Augusta.'" He went on to fill-in Jesse with additional information. "My wife has cooked for and fed many a miner. She ran a bakery when I was running my store. We came from Kansas in the Pikes Peak Gold Rush as man and wife, and chased from one gold settlement to another." Jesse and Tabor, both being Republicans, got along just fine.

Before leaving Leadville, Jesse was treated to some of Mrs. Tabor's cooking, which was a highlight of his trip. He remarked about the excellent pie and bread that she baked, which accompanied the meal.

Jesse left on the stagecoach for Kokomo, 20 miles to the north, where he planned another short stay. He had only one burden, that of his trunk filled practically full of specimens. It was very heavy and the stage company said it would have to go separately if they had a full load of passengers with all their usual luggage. However, Jesse was able to keep the trunk with him, for the coach was only half full. Nevertheless, his trunk made a full load for the stage. At Kokomo, Jesse arranged for his trunk to be freighted to Georgetown, but only after a few more specimens from the Kokomo area of the Ten-Mile District were added. These samples contained galena, zinc blende, iron pyrite and specular iron, and considerable amounts of lead and silver sulphurets.

Summit County, in which Ten-Mile was located, was a large county taking in all the area west of the Continental Divide clear to the Utah boundary, with Grand and Route on the north and Lake and Gunnison counties on the south.

The Ten-Mile District had been prospected by several different parties, but no high-grade ore was found in quantity. In the summer of 1878, George B. Robinson, a leading Leadville merchant, outfitted an old prospector named Charles Jones, and the Seventyeight, Smuggler and other mines of the Robinson group were found. At a later date, the Wheel of Fortune and Grand Union were discovered.

It was then that people started to move into the area to stake off claims, sometimes on top of the snow in late winter.

GEORGETOWN COURIER

THE MEYER SAMPLING WORKS and Harrison smelter are shown in this sketch, the first such operations in the Leadville Mining District.

EADVILLE

In this elevated region, snow falls deep and often, and there is usually five or six feet of it on the ground from January to late April, but nothing could stop the fever-heat of excitement that set into the Ten-Mile District. Prospectors kept coming in over routes that were thought to be impassable. Some stopped here on the way to Leadville. Trees were felled, cabins built and tents pitched on top of the snow. Men would seek for unclaimed ground, clear away the snow from a chosen locality and then sink a shaft in search of deposits or veins. This haphazard style of prospecting was occasionally successful, and a few good strikes were made on Sheep, Elk, and Jack mountains, all of which greatly advertised the fame of the Ten-Mile. Townsites were staked off for a distance of four miles down the valley.

Starting near the head of the Ten-Mile Valley, the mining camps were Carbonateville, Summit (or Ten-Mile), Robinson, Kokomo and Recen. There were from 30 to 50 arrivals daily all through the spring, when the melting snow made the poor dirt roads almost impassable.

With the opening of summer in 1879, Kokomo claimed a population of over 1,000 and had an organized city government, a bank, hotels, stores, saloons, sawmills and the telegraph, where there was not a single settler a year before. A newspaper and several smelters were in the new camp, but not yet in operation.

The Robinson Consolidated Mines comprised the following claims in the area, mostly consisting of 10 acres each (of surface ground): The Seventyeight, Little Giant, Big Giant, Undine, Smuggler, Pirate, Checkmate, Rhone, Ten-Mile, G.B.R., and a couple more. William E. Musgrave, general manager for the owner, was developing the entire property with shafts, drifts and levels, and had some 40 men at work.

This was considered the great mine of this section and good enough to give any district fame and reputation. The Seventyeight, Undine and Smuggler all showed the same large, high-grade deposit of mineral, there being one great incline vein, or lode, pitching below and nearly parallel with the surface of the mountainside. The thickness of this vein was from two to eight feet, averaging four feet, and most of it was composed of ore containing from $100 to $250 a ton. The second-class material assayed at 70 ounces of silver and was kept on hand for the smelter soon to be operating for the mine. Ore was transported to Leadville over terrible roads, ten miles on runners over the snow and seven miles in wagons.

The Wheel of Fortune mine was located on the summit of Sheep Mountain, overlooking the valleys of the Ten-Mile and Eagle. It comprised four claims: The Wheel of Fortune, Star of the West, Spare Ground and Highland Mary. Jacob Hecht discovered mineral nine feet from the surface in the summer of 1878. In time, an immense deposit of ore was opened and daily shipments were being made to Leadville during the winter and spring of 1878. The first 300 feet of drifts were excavated in a continuous deposit of mineral and crevice averaging six feet thick, and extremely large amounts of valuable mineral were developed. Twelve feet of crevice was discovered at one place. The first-class ore generally yielded from 80 to 125 ounces of silver per ton, and much of it carried considerable lead.

On the same mountain, the Carbonate Company had sunk a dozen shafts, some of which were reported to show mineral in large amounts. Among other mines were the Black Dragon and Invincible, Big Horn, Baby Mine, Little Fortune, Hidden Treasure, Tip Top, Rattler, Gray Eagle, Idalie, Ballarat, Anni Lisle, Roderick Dhu Ruby, Silver Tip and Paris.

GEORGETOWN COURIER

FRYER HILL mine buildings were sketched from Little Stray Horse Gulch for the "Georgetown Courier." Timbering for the Pittsburg, New Discovery and Winnemuck mines, as well as other mines, was cut from the heavy upland forest that once covered this hill—leaving only the tree stumps seen here.

W. H. JACKSON PHOTO
DAVID S. DIGERNESS COLLECTION

THE LEADVILLE MINING CAMP was booming in 1885, when the great William Henry Jackson toured the region.

Carbonate Hill—on the left, above the business district—was covered with prospect holes and mine dumps. The "Picture Maker of the Old West" recorded this view as he pointed his massive camera toward the southeast.

ALEX MARTIN PHOTO
RICHARD RONZIO COLLECTION

WASH DAY was every day for the woman taking in laundry in this view. Housing in the Carbonate Camp was so crowded that the laundry usually was done outdoors in good weather—sometimes beside the "family transportation." A shingle hung at the door possibly advertised the hard-working lady's laundry service.

On the main range was the Grand Union, or Lenon, mine, one of the first found and most valuable of the district. It was 12 feet wide, with ore carrying from a few ounces in silver up to 200 ounces per ton. Some very high assays were obtained and one sack of ore was sold at the rate of 50 cents a pound.

The Iron lode, near the Robinson, was a valuable galena vein, with the top ore converted into carbonates by decomposition a little more extensively than in most neighboring mines. There were many locations well up on the southeast face of this mountain, near timberline. Among them, the Idalie had a vein yielding from 8 to 311 ounces of silver per ton. During the winter, the Silver Tip was discovered by sinking a shaft at random on unclaimed ground. It showed eight feet of ore carrying from 20 to 100 ounces of silver per ton.

North of Sheep Mountain are the Elk and Jack mountains.

Near the mouth of Kokomo Gulch was the town of Kokomo. On Elk Mountain were the White Quail, Climax, Mountain Lode, Silver Chain, Elk and Governor Gilpin, the last-named mine yielding ore worth from $60 to $140 a ton. Two lodes were found in the town of Kokomo by men while excavating for cellars. One carried galena worth up to $200 a ton, and the other contained copper pyrites. On Jack Mountain were the Rising Sun and Racen lodes. The Frank lode was on Ford Mountain and carried 60-percent lead and 90 ounces of silver. On the eastern side of the district — on the west side of the Ten-Mile Range — the Colorado Springs Company had mines containing large amounts of ore.

The Eagle River starts from the vicinity of Tennessee Pass, west of the headwaters of the Ten-Mile, and flows northwesterly between the southern Gore Range and a more westerly range of mountains of the northern Sawatch Range into the Colorado River. It was the newest mining district of the almost-unexplored regions of western Colorado in Summit County. The mountains that enclose the river contained many silver veins, some of them assaying as high as 1,100 ounces. Many prospectors went in there, and in a beautiful park, Eagle City was located. West of the headwaters of the Eagle River is the Mount of the Holy Cross, whose eastern face always shows vast snowfields taking on the form of a cross. The snow fills two mammoth ravines, or crevices — one vertical and the smaller of the two, horizontal. The height of

W. H. JACKSON PHOTO
DAVID S. DIGERNESS COLLECTION

THE NEW ORO CITY was the second gold-mining camp in California Gulch. It formed two and one-half miles further up the gulch from the original Oro City. The Printer Boy gold lode was responsible for the establishment of this new settlement in the late 1860's.

MR. & MRS. ERNEST GERBAZ COLLECTION

THE MINING BOOM of the great Carbonate Camp was in full swing about 1879.

the cross is about 1,500 feet and the arms are each about 700 feet long. The natural cross, a landmark, is visible from north of Fremont Pass. Jesse noted in his journal that he first saw the great cross from Argentine Pass in the early 1870's. William H. Jackson took the first close-up photograph of it, and it was the subject of conversation when Jesse met Jackson for the first time.

Jesse was favorably impressed with the Ten-Mile Mining District and the town of Kokomo, where he stayed. He noted that the town was a marvel of activity, and he took many mental notes in addition to data he recorded in his journal, as is evidenced in later additional entries after the trip.

Jesse continued his return trip to Georgetown via stagecoach, hauled by what he called "a very spirited six-horse team." The coach went via the new Loveland Pass wagon road, a distance of 36 miles. The horses virtually flew along, and the coach rode on two wheels at times, almost turning over at one point high on the pass.

Jesse Randall returned to Leadville many times during the next few years gathering the news and collecting mineral specimens. On one of these visits — in 1887 — he met Horace A. W. Tabor, who was in Leadville attending to some business. Of course, Tabor owned the Matchless mine, which was fabulously rich in silver. Jesse also noted in his journal

RICHARD RONZIO COLLECTION

CHESTNUT STREET is shown here in a very early photograph, circa 1878. A six-ox team was hauling a freight wagon through town at the time. Notice the enthusiasm of the serious drinkers and the bartender.

GEORGE WAKELY PHOTO
RICHARD RONZIO COLLECTION

IN THIS VIEW OF 1880, California Gulch opened up, showing the early beginnings of Leadville. The main part of town is visible in this northwest viewpoint.

that wealth, over a number of years, had gone to Tabor's head and had made him a changed man. He spent money at a rate very close to, or even exceeding, his vast income. In Denver, he built the Tabor Block, a six-story stone building at Larimer and Sixteenth streets. This was followed by the five-story brick Tabor Grand Opera House and combination office building, an elegant structure at Sixteenth and Curtis streets. The Tabor Grand Opera House was extremely plush, having only the best of everything. The interior finishings were mostly of imported materials, such as mahogany, cherry wood, marble, crystal chandeliers, tapestries, etc., from such places as Honduras, Japan, Italy and France.

Tabor financed the construction of an opera house in Leadville, the Leadville Bank and the Tabor Grand Hotel. The opera house, he donated to the city. He owned considerable property in Leadville. St. George's Episcopal Church purchased the land that it occupies from Tabor. He cancelled the last $1,000 owed to him and donated the magnificent bell for its bell tower so the church could be debt free. He owned many enterprises, two lumber companies, a water company, a gas-light company, a horse-car (city-transportation) company, and a real-estate company. He incorporated the Tabor Milling and the Smelters Supply companies, and established the Leadville Stock & Mining Company, as well as others. In the Leadville mining area, he owned and had interests in many local mines. He invested in all types of ventures in the United States and abroad, speculating — sometimes wildly — in mines in the southern portion of The Mineral Belt. He started a smeltering business in Ashcroft, as well as incorporating a city-transportation street-railway system for the town of Aspen.

Early in 1883, Tabor divorced his wife, Augusta. She contested it, but finally had to give in to Tabor's demand. He was interested in another lady, the young and beautiful Elizabeth McCourt Doe, daughter of Peter McCourt, who became known in Central City by the miners as "Baby Doe." She had divorced her previous husband, who had failed to hit it rich in a Central City mine after moving there from Chicago.

The divorcee then moved to Leadville, where Tabor met her. Tabor and Baby Doe were later married in Washington, D. C., while Tabor filled out a short unexpired term in the U. S. Senate. A republican of deep-seated convictions, he had entered the political arena and was elected Lieutenant Governor in 1880. In January of 1883, the Colorado General Assembly appointed Tabor to serve the unexpired part of the Teller term which ended March 4, 1883. He was appointed to this post after losing an election for political office to a Democrat. Senator Tabor had faithfully served his party in many capacities: as chairman of its state central committee; as chairman or member of its state executive committee; and as a counselor, guide and friend in many seasons of political

W. H. JACKSON PHOTO
RICHARD RONZIO COLLECTION

THE VERY BEGINNINGS of Leadville were recorded on glass in this view of 1880. W. H. Jackson captured this scene from Carbonate Hill, as he looked toward the north.

strife. He gave generously to his party in the days of his prosperity, and after adversity came upon him, he freely continued to give it his services and influence. However, the wide publicity given the divorce of Horace and Augusta Tabor caused Washington and Denver socialites to ostracize Baby Doe, and the resulting and far-reaching scandal hurt Tabor's political career. His popularity dropped. The Republican Party would no longer back him after previously having given him vigorous party support. Tabor had given large contributions to the party. In addition to this harmful publicity, Tabor was mocked by poet Eugene Field in the *Denver Tribune*. In all, this ended his political life.

The newly-wed couple made Denver their home and Tabor provided his devoted wife with a mansion in which to live. He gave her magnificent jewelry and other luxury items such as carriages and horses to match her elegant dresses, and always arranged to have flowers surrounding her box seat when she attended performances at the Tabor Grand Opera House. Her father, Peter McCourt, was given the job of general manager of the opera house.

Two daughters had been born to Horace and Baby Doe by the time of Jesse's visit. The first was named Lillie and the second was Rose Mary Echo. A short time later, William Jennings Bryan nicknamed the Tabor's second daughter "Silver Dollar" because he thought her laughter had the ring of a silver dollar, and thereafter she was known as Silver Dollar Tabor.

To add an extra-special and colorful touch to the grounds surrounding his mansion, H. A. W. invested in a large flock of peacocks to run free over the property, which already featured much ornate statuary.

His former wife had been shocked at his display of affluence, and was greatly distressed by his gambling and philandering. At the time of Jesse's visit, he was starting to feel the bad effects of this practice. Some of his mines had become exhausted, a number of his far-flung investments proved unsound, and he was suffering heavy losses from gambling. Still further problems added to his misfortunes; claims against many of his mines involved him in one costly litigation after another. He mentioned that he had mortgaged some of his property to help finance the mounting debts.

Jesse noted that H. A. W. assured him that this was only temporary and that he intended to continue his usual life style of extravagance. Money had gone to Tabor's head. He loved his title as the Silver King and he intended to live like one.

As time passed, Tabor continued to lose his fortune by practically giving it away and had nothing to fall back on when silver was demonetized in 1893.

Tabor did much for Leadville by building a system of waterworks and a telephone exchange, and was engaged in many other enterprises. His investments in Denver tremendously stimulated improvements in the city. His influence and assistance inspired everyone with energy and enthusiasm. He was investing all of his great wealth in the development and improvement of the city and state. He spread faith and confidence in both wherever he went and in all that he did.

When the Silver Panic of 1893 hit, the devastating financial storm nearly ruined the state's mining industry and wrecked many Colorado fortunes, including Tabor's. The old pioneer, H. A. W., was again beset by hardship, as he was when he

W. H. JACKSON PHOTO
RICHARD RONZIO COLLECTION

DURING THE GOLD RUSH — and for many years thereafter — well-organized shopkeepers displayed their goods out in front of their store buildings. As could be expected, the almost-daily occurrence of showers at high elevations made for a hasty retreat of display goods from these open-air markets.

DAVID S. DIGERNESS COLLECTION

THE CARRIAGE BUSINESS was very lucrative during the 1880's in the Carbonate Camp. The large number of people immigrating to the mining district usually came without vehicles of their own, and soon realized that they needed to become owners of horses and rigs.

STATE HISTORICAL SOCIETY OF COLORADO

THE MOUNT MASSIVE HOTEL was located near California Gulch, and afforded clean beds, bathing facilities and food to weary travelers and miners of 1878. Many prospectors took advantage of this luxury to have a bath and a short recovery period from tramping over countless rocky hillsides.

GEORGETOWN COURIER

THE GRANT SMELTER was built at Leadville in 1878, one of a dozen smelters in the area. However, it was destroyed by fire in May of 1882.

DENVER PUBLIC LIBRARY, WESTERN HISTORY DEPARTMENT

ONE OF LEADVILLE'S MANY SMELTERS, shown here in the 1870's, was hastily built of wood to handle the onslaught of ore from this rich Mineral Belt area. Large stacks of 100-pound lead bars containing silver were stacked outdoors for shipment to the nearest railhead. Smelters operating in this area during this period included the Grant, Harrison, Malta, Adelaide, Berdell & Witherell's La Plata, Germania, Iowa, Garbutt, Saint Joseph and Little Chief.

THE ROBERT E. LEE MINE pithead was one of the larger structures to be built on Carbonate Hill during the boom year of 1880.

W. H. JACKSON PHOTO RICHARD RONZIO COLLECTION

W. H. JACKSON PHOTO
RICHARD RONZIO COLLECTION

THE ARTISTIC SKILLS of Western photographer W. H. Jackson were put to use to produce the full-color effect of this early lithographic print. Lead-silver bullion was being produced in great quantities in the Leadville Mining District, so great that this smelter used an 18-inch-gauge rail system within its plant to transport the precious metal to shipping and storage areas.

A. T. PIERCE PHOTO
DAVID S. DIGERNESS COLLECTION

TYPICAL OF THE SMALL MINES near Leadville, this operation had a steam hoist, which raised and lowered the ore buckets in the vertical shaft. The small size of the dump indicates that this is a relatively new venture. The buildings were: privy, coal shed, tool shed, stable and bunkhouse—besides the pithead itself.

HARRISON REDUCTION WORKS.

DAVID S. DIGERNESS COLLECTION

PROSPECTING NEAR LEADVILLE.

DAVID S. DIGERNESS COLLECTION

GEORGE WAKELY PHOTO
RICHARD RONZIO COLLECTION

DURING THE EARLY DAYS the main business street of the "Cloud City" ran east and west—near California Gulch. Notice the large number of trees on Carbonate Hill, as well as the many cabins and tents.

DAVID S. DIGERNESS COLLECTION

CHARLES WEITFLE PHOTO
RICHARD RONZIO COLLECTION

THE TOUGHEST FELLOWS one could meet in pre-automobile days were the blacksmiths of any given town. Beating iron all day long gave blacksmiths arms of "iron." This group provided the photographer their best smiles in the Leadville of 1879.

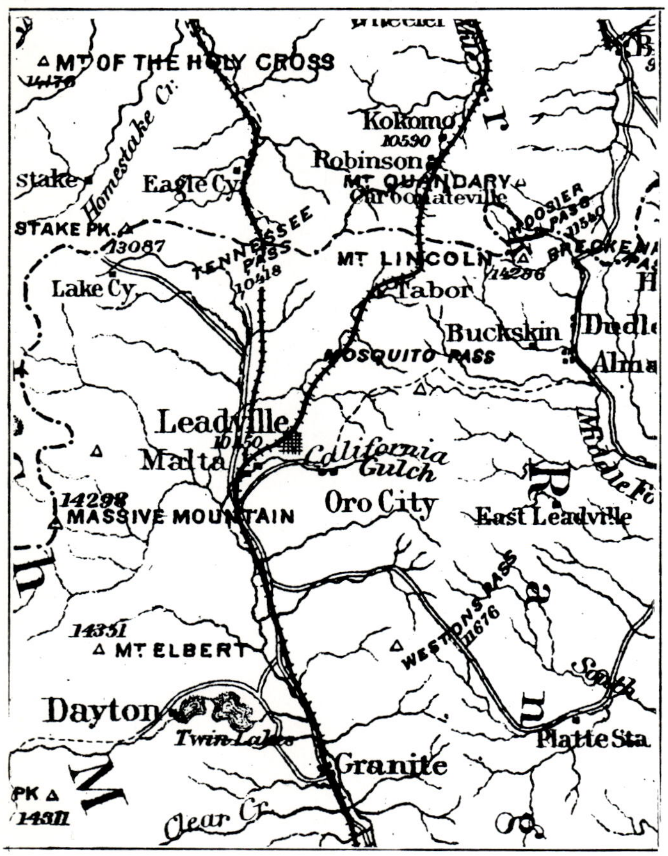

GEORGE D. WAKELY PHOTO
STATE HISTORICAL SOCIETY OF COLORADO

SOMETHING FOR EVERYONE was offered at the corner of Harrison Avenue and Chestnut Street (now First Street) in 1879. What looks like a Las Vegas-type chapel adorned the foremost intersection in Leadville, with a gift shop or fruit-and-vegetable stand attached on Harrison Avenue. It is interesting to see how the shopkeepers displayed their wares each day on the board sidewalk.

first came to The Mineral Belt.

In 1898, President McKinley appointed Tabor postmaster of Denver, the position he held at the time of his unexpected death, which occurred on the 10th of April, 1899. H. A. W. Tabor was a frank, honest, open-hearted man — an unusual combination of gentleness and ruggedness. Despite his obvious flaws, no more generous a man, none truer to his friends nor more responsive to their wants in times of need, was found than H. A. W. Tabor.

Economic and Labor Troubles at Leadville

During the serious economic depression brought on by the Silver Panic of 1893, Colorado had more than its share of labor problems. One of the worst events took place in the Leadville Mining District during 1896-1897.

Volume II of *The Mineral Belt* will delve into this troubled

STATE HISTORICAL SOCIETY OF COLORADO

AMONG THE WELTER of false-fronted buildings, gaudy signs and overhanging balconies in early-day Leadville, it is interesting to note that one John Lieninger operated a book and stationery store.

period in greater detail than we will attempt to do at this time; however, we feel that a brief exposition on the "Great Leadville Miners' Strike" is in order here.

The peak year for silver production in the Leadville Mining District was 1880. At that time, there were 14 smelters and ore-reduction plants in the district, and the output of silver topped $11-million—although the big mines held production around $10-million for several more years.

When it was disclosed that certain of the district's mines had been recklessly exploited, and their owners had borrowed heavily in order to pay stock dividends, silver-mine stocks dropped in value from many dollars per share to only a few cents. The *exposé* further revealed what appeared to be inside manipulation of mining company stocks by unethical capitalists. As a result, investors became panic stricken and wildly "dumped" their shares—thereby breaking the market. One-by-one, banks in The Mineral Belt failed and were forced to close their doors. Meanwhile, once-boisterous capitalists and other "sporting dandies" discovered they had urgent business to attend to elsewhere.

During this same period, Leadville's Grant smelter burned to the ground, and several other important buildings in the city met the same fate, including hotels and department stores. Population began to recede and silver production declined. However, Leadville was hardly prepared for even-worse times to come. In 1893 the mints of India stopped buying silver for coinage, and the U.S. Congress shortsightedly repealed the Sherman Silver Purchase Act, thus ending bimetallism and undercutting the economic base of Colorado and other Western silver-producing states. This asinine action by the Federal government brought on the terrible Silver Panic of 1893, causing the price of silver to fall to such a low level that many mining companies could no longer afford to mine the once-precious grey ore.

Reviewing the events of 1893, the *Leadville Herald-Democrat* declared "Ruin and bankruptcy stared every mining man, every smelting man, and every business man in the face."

OVERLEAF: The main business street of Leadville in 1879 was Chestnut Street, which ran east and west. Of special interest in this scene is the assortment of signs displayed to lure prospective customers into the shops.

G. D. WAKELY PHOTO
RICHARD RONZIO COLLECTION

TMARKET
SHOE STORE
SMITH
TOBACCO
LLE.

BAKERY
BOOK STORE

STATE HISTORICAL SOCIETY OF COLORADO

PAPA AND BILLY KELLER built the Grand Pacific Hotel in 1879, on this interesting block in Leadville. The Leadville Beer Garden was constructed with three entrance doors in the front. The two outer doors probably were private shop entrances.

THE SEEMINGLY UBIQUITOUS BURRO made prospecting and mining possible in many parts of Colorado. The Leadville Minging District was no exception, and the scene below portrays a packtrain on one of the "Cloud City's" downtown streets, ready to head for the mines.

STATE HISTORICAL SOCIETY OF COLORADO

W. H. JACKSON PHOTO
STATE HISTORICAL SOCIETY OF COLORADO

COLOR PHOTOGRAPHY did not exist during the 1880's, when this photograph was taken; however, W. H. Jackson put his ability to work to bring color views within the average person's means. This full-color lithographic print represents Colorado hydraulic mining in action.

The fantastic silver boom was over, and in its place, panic stalked the trails of The Mineral Belt.

Thousands of hardrock miners were thrown out of work as a result of this clearly unavoidable but none-the-less severe depression. And those miners who remained at work in the pits felt very insecure, to say the least. During September of 1893, Leadville district miners signed an agreement with the mine owners which in effect reduced wages from $3.00 to $2.50 per day.

The general dissatisfaction of the miners festered for nearly three years. Then, during June of 1896, some 1,600 members of the Western Federation of Miners walked off their jobs in the Leadville district—closing 90 percent of the mines.

When the mine owners brought in non-union men, violence was threatened by Western Federation men. As a result, virtually no mining was done until September 20, 1896—and

WATER FLUMES were constructed to carry water over gulches in the mining areas. The flumes made hydraulic mining possible in placer diggings.

GEORGETOWN COURIER

STATE HISTORICAL SOCIETY OF COLORADO

THE CLARENDON HOTEL, located next door to the Tabor Opera House, was one of Leadville's largest hotels. Built before brick was readily available, this hotel was a typical wooden Victorian building of the era. Shopkeepers rented out space on the street level on each side of the entrance to the hotel (covered by an exterior cold-weather foyer). Next door was the Atchison, Topeka & Santa Fe ticket office, offering connections via the railroads serving Leadville.

THE CARBONATE HILL SCHOOL taught many of Leadville's children their "three R's." Many people feel that old-time local schoolhouses such as this one provided a better education than modern school systems.

STATE HISTORICAL SOCIETY OF COLORADO

OPPOSITE, ABOVE and BELOW: The Pendery & Company Mining Bureau occupied the same new brick building as the ticket agency and freight office of the Atchison, Topeka & Santa Fe Railway. This side street offered patrons assorted services—a barber shop, tailor shop, assay office, saddle shop and a sales outlet for "Hand Made Sour Mash Whiskey" from the Paris Bourbon Company of Kentucky. The ornate structure (below) was Leadville's Central School, as it appeared in 1884. The architect took great pride in the design of this building, which was set among the wooden shacks of the town.

STATE HISTORICAL SOCIETY OF COLORADO

W. H. JACKSON PHOTO
RICHARD RONZIO COLLECTION

DAVID S. DIGERNESS PHOTO

THE MINE DUMPS of Carbonate Hill still show the remains of a bygone age above the city of Leadville. Late-afternoon sunshine warmed the dwellings of Leadville — painted in an array of color quite unlike the old mining days when nearly all buildings were white, or went unpainted.

WINTER SNOWS at times piled up so deep that the only means of transportation over wagon roads was by sled. This view of about 1910 looks out over the Arkansas Valley, with Mount Massive in the distance.

GEORGE L. BEAM PHOTO
DAVID S. DIGERNESS COLLECTION

OPPOSITE: The Emmet Shaft at Leadville is a typical example of the many relatively small pitheads that dotted the hillsides of The Mineral Belt.

DAVID S. DIGERNESS PHOTO

STATE HISTORICAL SOCIETY OF COLORADO

THE GRAND CENTRAL THEATRE was one of many early-day structures to succumb to fire during Leadville's hectic history.

GEORGETOWN COURIER

THE CLARENDON HOTEL was a popular establishment during the 1880's. It was located on Harrison Avenue in Leadville, adjacent to the Tabor Opera House.

W. O. LUKE PHOTO
DENVER PUBLIC LIBRARY,
WESTERN HISTORY DEPARTMENT

THE MEAT MARKET of Pearsall offered a wide selection of cuts. The patron could choose such mouth-watering delicacies as pork, veal, chicken and duck. If this were a present-day occurrence, the government health inspector would be appalled by the sight of this display in the open air of Leadville.

OPERA BILLS adorn the false front of the Woods Opera House in Leadville. This was before the turn of the century, when gas lamps adorned the streets and it was quite fashionable for gentlemen to take their wives or lady friends to the opera. At the time this photograph was recorded, Chestnut Street (First Street) was the main business street.

STATE HISTORICAL SOCIETY OF COLORADO

even the mines' water pumps were shut off, causing the flooding of many of the pits. On September 20th, the Colorado Militia was called out to restore peace in the district, and mining was resumed on a limited scale shortly thereafter.

With the continued importation of non-union men and the presence of State troopers—in effect backing the mine owners—the Western Federation of Miners was forced to give up their efforts, and on March 9, 1897, the strike was called off.

The effects of the strike—combined with the depressed price for silver—very nearly wiped out the mining industry in the district. And a number of the more marginal mines were closed permanently due to flooding and silting while the pumps were off.

DAVID S. DIGERNESS PHOTO

NOT TO BE OUTDONE by other members of the cloth, the Episcopalians of Leadville built their house of worship in a respectable design to entice even the worst of the weekday sinners through their doors. The bell was donated by H. A. W. Tabor, who also held the mortgage, which, in turn, was paid by Tabor. St. George's Episcopal Church is a landmark of the district.

Even without the Great Leadville Miners' Strike, it is doubtful that silver mining could have remained a viable industry. The insensate Congressional action ending bimetallism very nearly ruined the economy of Colorado. Even more shocking was the fact that Congress and the Administration did absolutely nothing to alleviate the monstrous problems they had caused. But then, the Federal government often has looked upon Colorado and the other Rocky Mountain states as though they are little more than colonies—not quite full-fledged states of the Union. Evidently, Washington thinks that these "colonies" are to be exploited at the whim of government, in league with monopolistic big business and the notoriously unscrupulous Eastern banking interests.

To this day, precious-mineral mining has not fully recovered in The Mineral Belt, and ghost mines dot the hillsides in many parts of the region—monuments to the colossal callousness of an insensitive National government—too far removed from the people of the West, both in time and space, to care what happened to them.

GEORGETOWN COURIER

THIS SOUVENIR of the Iron-Silver mine on Iron Hill was produced during the heyday of hardrock mining in the Leadville district.

W. H. JACKSON PHOTO
STATE HISTORICAL SOCIETY OF COLORADO

ST. GEORGE'S Episcopal Church in Leadville was decorated for the Christmas holidays, as this unusual interior view shows.

HARRISON AVENUE — the main thoroughfare of Leadville — was filled with the sounds of horses and sled runners in the winter of 1897. No matter how deep the snow, the city continued its fast pace.

STATE HISTORICAL SOCIETY OF COLORADO

THE LEADVILLE COURTHOUSE for Lake County (at right) crowded against the more conventional mercantile building next door. According to business-directory list-

ROBERT M. DAVIS PHOTO
DAVID S. DIGERNESS COLLECTION

ings for photographer Brisbois—who occupied the second floor of the business building—this view must have been taken during the 1880's.

DAVID S. DIGERNESS COLLECTION

LEADVILLE'S SMELTERING INDUSTRY became the state's second greatest activity of this type—surpassed only by Denver's great smelters. The first smelter in the Carbonate Camp made its appearance in October of 1877, when the St. Louis Smelting & Refining Company fired-up a blast furnace in what they called the Harrison Reduction Works. Several of the Harrison staff are shown here on what may have been the porch of the Harrison office building.

THE OFFICE of the Arkansas Valley Smelting Company was a log building in 1898. Company officials were sitting for their portrait. In later years, this company became the largest smelting operation in the Leadville Mining District, and still later, it was absorbed by and became known as the American Smeltering & Refining Company. The employees of this smelter lived in Stringtown on the southwest edge of Leadville, outside the city limits.

DAVID S. DIGERNESS COLLECTION

EDWARD N. CLEMENTS PHOTO
RICHARD RONZIO COLLECTION

THE ST. LOUIS HOUSE was a popular night spot for Leadville's miners to "drink their fill"—and perhaps enjoy the "girls of the night," who lived upstairs. A meat market and the Union Pacific Saloon were next door. It was said that during the 1880's the more-elite social drinkers of Chestnut Street attended the U.P. Saloon, while the serious drinkers quenched their thirst at the St. Louis House.

CHESTNUT STREET, as viewed from Harrison Avenue, appeared to be the businessman's dream come true in 1887. Building sites that originally could have been purchased for $25 were selling for thousands of dollars by early 1879. Notice that electric utility poles have made their appearance.

W. H. JACKSON PHOTO
DAVID S. DIGERNESS COLLECTION

CHESTNUT STREET. LEADVILLE.

HARRISON AVENUE gained swiftly in popularity as Leadville's main business street as merchants set up shop there in the late 1880's. This was partly due to the fact that Chestnut Street ran up and down hill, while Harrison Avenue was relatively level. Notice the apparent lack of saloon signs.

LUKE & WHEELER PHOTO
RICHARD RONZIO COLLECTION

HARRISON AVE. LEADVILLE.

DAVID S. DIGERNESS COLLECTION

DURING THE LATE '30's, Leadville's Harrison Avenue was typical of towns of that period, with gasoline pumps mounted on sidewalks in front of business blocks. In those days, a concrete highway was still a luxury, so only the main part of the avenue was paved, while the parking space by the curb was still gravel.

HARRISON AVE. LOOKING NORTH FROM 3RD.

DAVID S. DIGERNESS COLLECTION

DAVID S. DIGERNESS PHOTO

MOUNT ELBERT, Colorado's highest peak—at 14,433 feet above sea level—rises above the sparkling waters of Half Moon Creek. It is late summer in the high country near Leadville and the air is crisp and clear.

LAKE COUNTY'S stately courthouse was the subject of this scene of about 1910. The Colorado & Southern Railway ticket office occupied the corner building (to the right) on the main thoroughfare of the city, Harrison Avenue. This was a handy place to purchase tickets since this location was near the center of town. Although the

C&S was proclaiming itself "The Colorado Road" at this time, the route to Leadville still was called the "South Park Line," as the sign on the building boldly states.

DAVID S. DIGERNESS COLLECTION

DAVID S. DIGERNESS PHOTO

THE MATCHLESS MINE without doubt is the most famous silver mine in all of Colorado. Shown in this recent color photograph, the mine stands as a landmark above the city of Leadville.

THIS ARTIST'S RENDITION of underground workings of a mine at Leadville shows timbers shoring-up the walls and roof to prevent cave-ins. To the right, the vertical bucket shaft can be seen.

GEORGETOWN COURIER

RICHARD RONZIO COLLECTION

ALMOST EVERY TOWN has had its "great snow," and some can even claim several which vie for the distinction. Here is some of the evidence of one of the Leadville entries—the great storm of March, 1899. This scene is between Fourth and Fifth streets, on Harrison Avenue.

STATE STREET. LEADVILLE.

DAVID S. DIGERNESS COLLECTION

STATE HISTORICAL SOCIETY OF COLORADO

A WOODEN PIPELINE was under construction when this scene was recorded in the 1890's. The Leadville district's mining and milling operations were increasing—and so was the need for water.

DAVID S. DIGERNESS COLLECTION

STATE HISTORICAL SOCIETY OF COLORADO

THIS INTERIOR VIEW of the Carbonate National Bank at Leadville portrays the tellers and bank officers of the period, about 1890.

DAVID S. DIGERNESS COLLECTION

DAVID S. DIGERNESS COLLECTION

THE TABOR GRAND HOTEL—at the northwest corner of Harrison Avenue and Seventh Street in Leadville—was one of the many enterprises H. A. W. Tabor invested his fortune into. This fine four-story brick structure was built during the 1880's. In the style of the Victorian period, it featured a false mansard roof decorated with a fancy cupola and iron-work along the roof edge. The hotel was opened in 1885 as the Tabor Grand and was renamed the Vendome in 1894. Only recently was the original name restored.

DAVID S. DIGERNESS COLLECTION

HOTEL KITCHEN.

H. A. W. TABOR AND HIS LEADVILLE ENTERPRISES

DAVID S. DIGERNESS PHOTO

THE ELEGANT Tabor Grand Hotel is shown here as it looks today—still in use—standing as a monument to the "Silver King of Leadville."

GEORGETOWN COURIER

THE TABORS arrived in California Gulch in 1860, and shortly thereafter, established a shop to provide meat and other necessities to the more than 5,000 men that stampeded into the area in search of riches. The name of the gulch came about when a '49er from California let out a yell on the headwaters of the Arkansas River, ''Gold! —— I have all of California in my pan!''

GEORGETOWN COURIER

H. A. W. TABOR

DAVID S. DIGERNESS COLLECTION

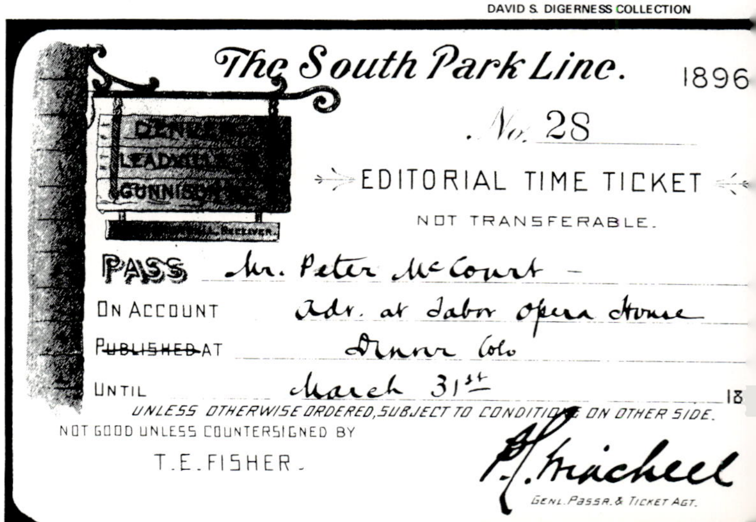

The South Park Line. 1896

No. 28

EDITORIAL TIME TICKET

NOT TRANSFERABLE.

PASS Mr. Peter McCourt –

ON ACCOUNT Adv. at Tabor Opera House

~~PUBLISHED~~ AT Denver Colo

UNTIL March 31st 18

UNLESS OTHERWISE ORDERED, SUBJECT TO CONDITIONS ON OTHER SIDE.

NOT GOOD UNLESS COUNTERSIGNED BY

T. E. FISHER.

GENL. PASSR. & TICKET AGT.

DENVER PUBLIC LIBRARY —
WESTERN HISTORY DEPARTMENT

AUGUSTA LOUISE PIERCE became Horace A. W. Tabor's first wife. This portrait was taken when Tabor was the Lieutenant-governor of Colorado.

DENVER PUBLIC LIBRARY,
WESTERN HISTORY DEPARTMENT

ELIZABETH McCOURT DOE, the daughter of Peter McCourt, became known in Central City as "Baby Doe" by the miners. She had earlier divorced her husband in Central City due to his "despicable failure" to strike it rich. She then moved on to Leadville where Horace "Silver King" Tabor — who had already struck it rich — met her. Horace and Baby Doe were later married in Washington, D. C.

GEORGETOWN COURIER

H. A. W. TABOR'S BANK, the Bank of Leadville.

A CONSIDERABLE AMOUNT of W. A. H. Tabor's Victorian taste still exists in Leadville's Tabor Opera House. This is one of the few remaining monuments created by this man who once had tremendous wealth and political power—only to lose it all in the Silver Panic of 1893. Thanks to the foresighted citizens of Leadville, this structure endures, still intact in an age when the wanton, senseless destruction of Colorado's great historic buildings is commonplace.

THE BANK OF LEADVILLE was heavily adorned with Victorian style brickwork — very expensive for 1878, when this view was recorded. Brick had to be shipped in at an outrageous cost, since the railhead had not yet reached Leadville from Denver when this building was constructed.

STATE HISTORICAL SOCIETY OF COLORADO

STATE HISTORICAL SOCIETY OF COLORADO

THE TABOR OPERA HOUSE still stands on Harrison Avenue in Leadville—once one of the most elaborate opera houses in Colorado. A covered walkway was built between the opera house and the Clarendon Hotel, providing patrons a passageway which protected them from the elements. Space was rented to shopkeepers on the street level, which helped defray the cost of this extravagant building.

TABOR OPERA HOUSE. LEADVILLE. CLARENDON HOTEL.

STATE HISTORICAL SOCIETY OF COLORADO

THE LEADVILLE HOME OF H. A. W. TABOR was but a small Victorian cottage patterned after many in the city. For many years — after the Silver Panic of 1893 — this house, as well as most other Leadville buildings, fell into disrepair. Today, however, the house has been restored and is open to the public, providing visitors a glimpse into the great silver-mining past of Leadville.

THE FINAL HOME of Baby Doe Tabor was this rough-cut cabin on the outskirts of Leadville, where she kept her vigil at the Matchless mine. The inset photograph was taken of her when she was a socialite in 1884; she died in her cabin in 1934.

DAVID S. DIGERNESS COLLECTION

G. B. SANBORN PHOTO
DAVID S. DIGERNESS COLLECTION

AN OVERALL VIEW of H. A. W. Tabor's fabulously rich Matchless mine is shown in the view of Leadville above.

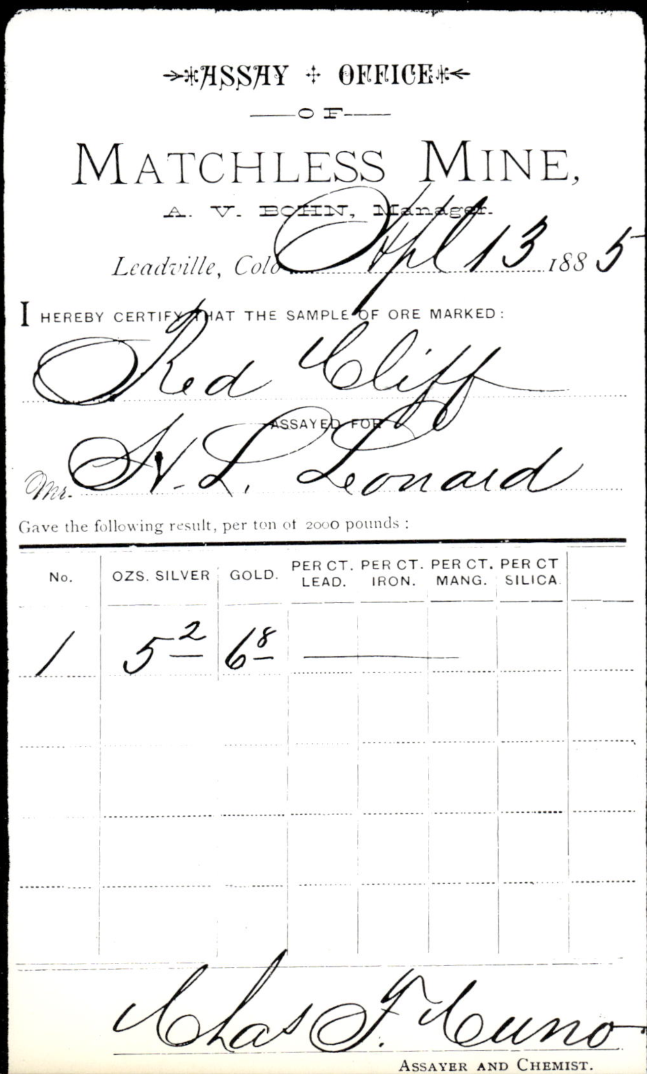

ASSAY OFFICE

—OF—

MATCHLESS MINE,

~~A. V. BOHN, Manager.~~

Leadville, Colo Apl 13 1885

I HEREBY CERTIFY THAT THE SAMPLE OF ORE MARKED:

Red Cliff

ASSAYED FOR

Mr. H. L. Lonard

Gave the following result, per ton of 2000 pounds:

No.	OZS. SILVER	GOLD.	PER CT. LEAD.	PER CT. IRON.	PER CT. MANG.	PER CT SILICA.	
1	5.2	6.8	—				

Chas F. Cuno

ASSAYER AND CHEMIST.

MORRIS ABBOTT COLLECTION — COURTESY OF WILLIAM R. JONES

1. THE LITTLE PITTSBURG. 2. NOSE PAINT SALOON.
3. STRUCK IT RICH. 4. STEADFAST.

DAVID S. DIGERNESS COLLECTION

STATE HISTORICAL SOCIETY OF COLORADO

JAMES W. JACKSON, iron founder and machinist, built this double-cylinder link-motion geared hoisting steam engine with 12-inch bore, 18-inch cylinders. The Little Pittsburg Consolidated Mining Company had four stationary steam engines similar to this one freighted to Leadville and put into operation during the spring of 1879. These engines went to work in their New Discovery, Little Pittsburg, Dives and Winnemuck mines.

GEORGETOWN COURIER readers were presented with this sketch of the Little Pittsburg Consolidated Mining Company at Leadville.

GEORGETOWN COURIER

MINING IN THE CALIFORNIA GULCH AREA SOUTHWEST OF LEADVILLE

DAVID S. DIGERNESS COLLECTION

CALIFORNIA GULCH was crossed by the Colorado Midland Railroad mainline in the city of Leadville, as this view (looking northeast) clearly shows. The Carbonate Hill School was located near the tall brick chimney in the center of this photograph — which was taken about 1910. The trackage leading off the left-hand side of the picture belonged to the D&RG's Iron-Silver Branch and sidings here served the Lanyon Zinc Company. Near the upper end of the trestle (to the right) the CM ran a connecting track to the C&S Mineral Belt line.

GRANT SMELTER. LEADVILLE.

DAVID S. DIGERNESS COLLECTION

G. B. SANBORN PHOTO
DAVID S. DIGERNESS COLLECTION

SMOKE from the Arkansas Valley Smelter nearly obscured Mount Massive in the background of this view.

MYERS SAMPLER & ST. LOUIS SMELTER.

DAVID S. DIGERNESS COLLECTION

GEORGETOWN COURIER ENGRAVING

PLACER DIGGINGS and sluicing for gold in California Gulch was done by using the limited water supply over and over again by each successive claim owner.

DAVID S. DIGERNESS COLLECTION

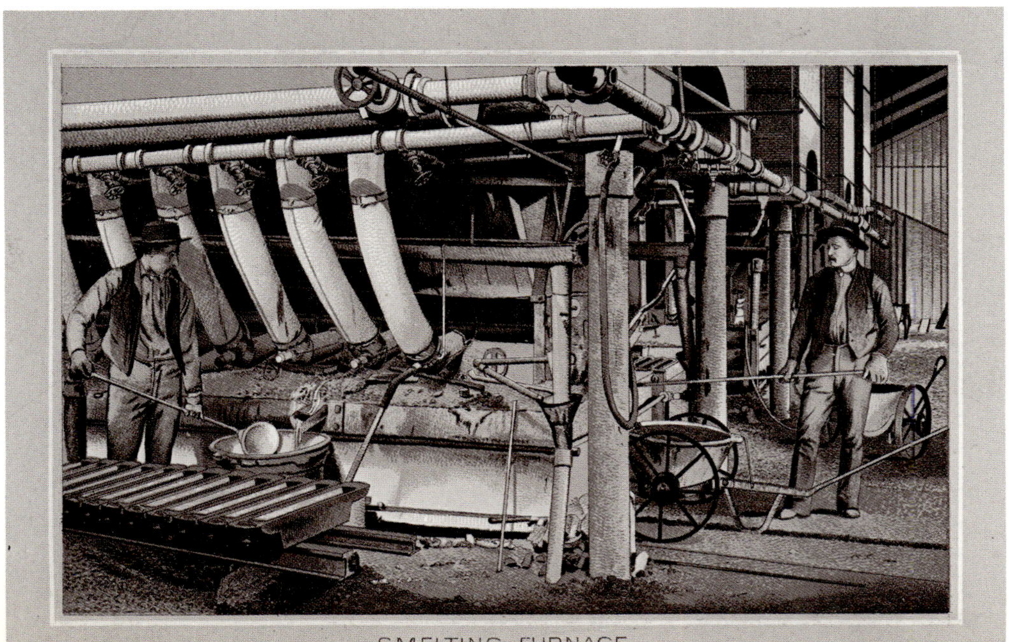

SMELTING FURNACE.

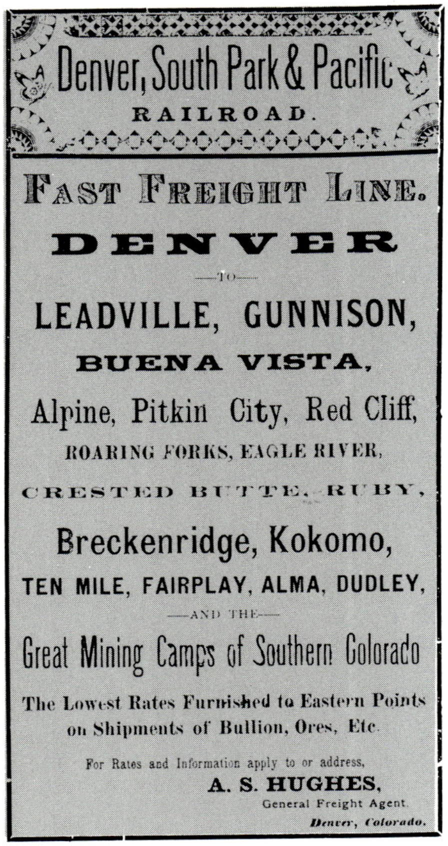

Denver, South Park & Pacific
RAILROAD.

FAST FREIGHT LINE.

DENVER

—TO—

LEADVILLE, GUNNISON,

BUENA VISTA,

Alpine, Pitkin City, Red Cliff,

ROARING FORKS, EAGLE RIVER,

CRESTED BUTTE, RUBY,

Breckenridge, Kokomo,

TEN MILE, FAIRPLAY, ALMA, DUDLEY,

—AND THE—

Great Mining Camps of Southern Colorado

The Lowest Rates Furnished to Eastern Points on Shipments of Bullion, Ores, Etc.

For Rates and Information apply to or address,

A. S. HUGHES,
General Freight Agent.
Denver, Colorado.

THE SOUTH END of Leadville is flanked by the Sawatch Range, which includes the two highest peaks in Colorado, Mount Elbert and Mount Massive. Leadville's smelters were surrounded by an array of trackage — coming from the mining ventures scattered all over the hillsides above town. The photographer was looking across the "Cloud City" from the west slope of Carbonate Hill.

DAVID S. DIGERNESS COLLECTION

RICHARD RONZIO COLLECTION

THE ARKANSAS VALLEY SMELTER was built in 1882. It had a capacity of 180 tons per day. During a merger of 1899, it became part of the American Smelting & Refining Company. Notice the high fence built around the property to keep out the public. This great old smelter closed on December 31, 1959.

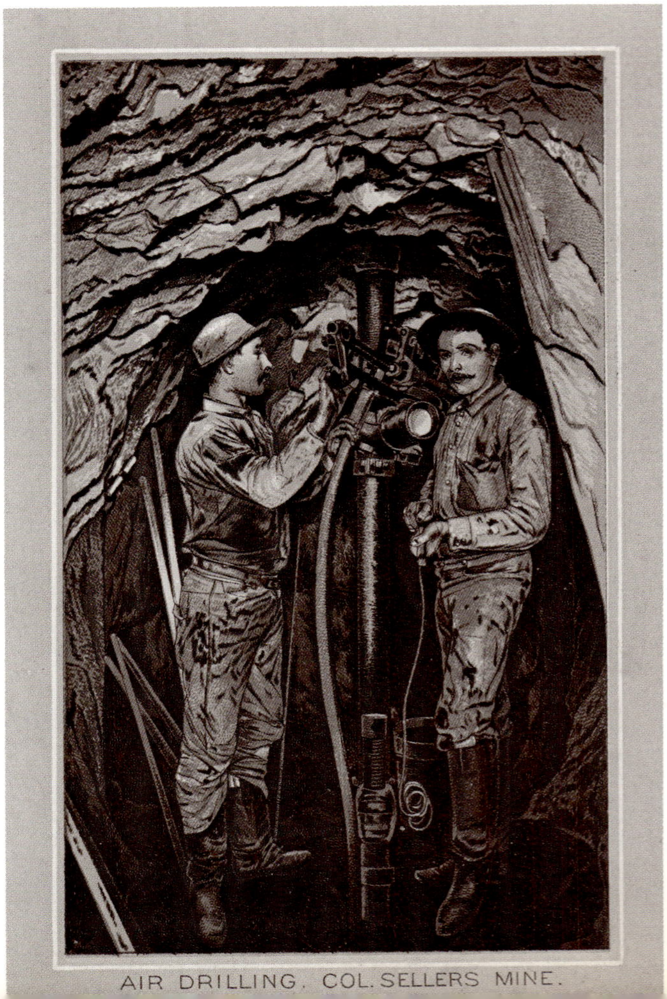

AIR DRILLING. COL. SELLERS MINE.

DAVID S. DIGERNESS COLLECTION

THE SOUTH PARK LINE

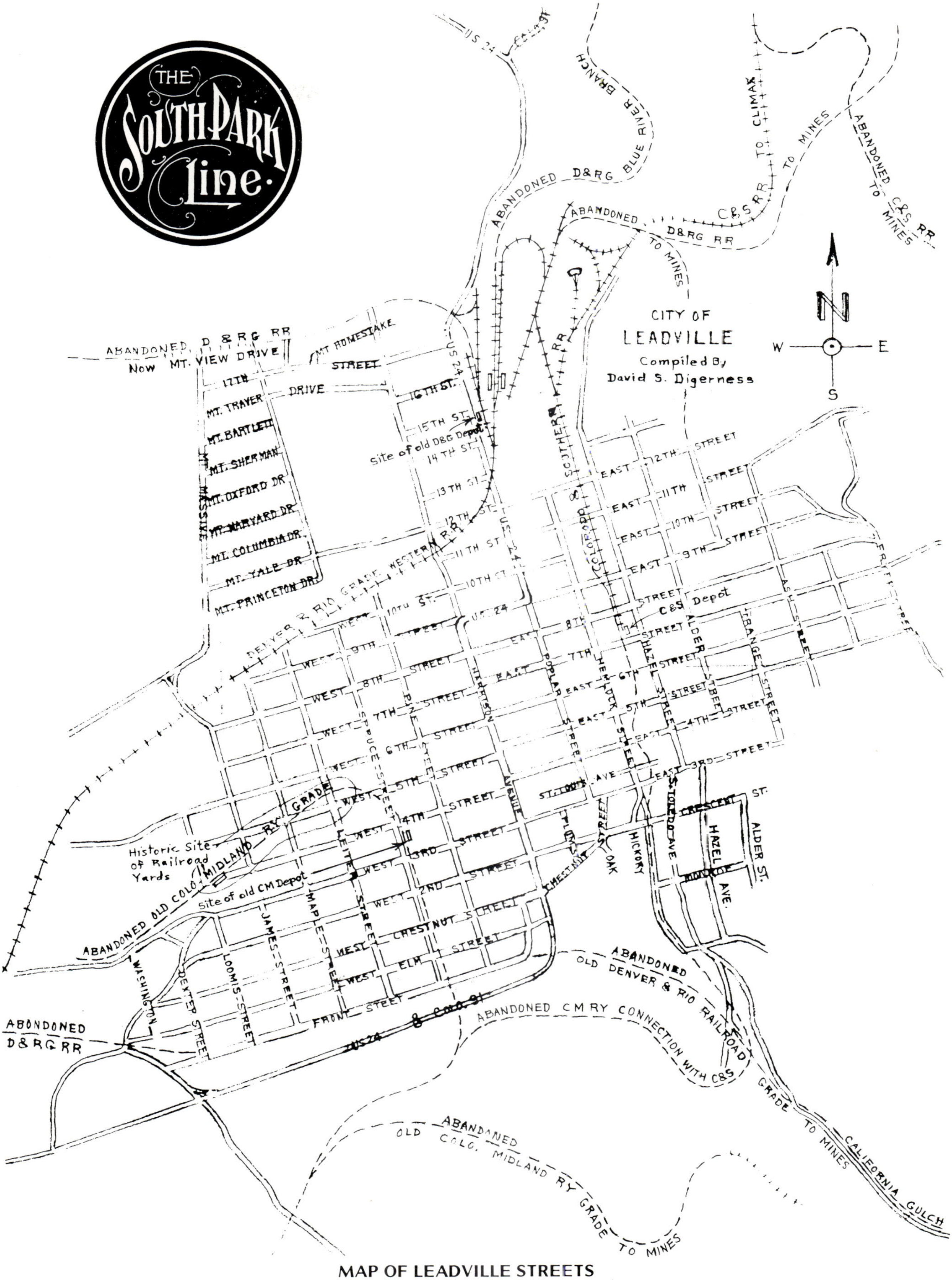

MAP OF LEADVILLE STREETS

RICHARD RONZIO COLLECTION

THE OFFICE of the Yak mine was located next to the power plant for the mine. Generators were fueled with coal, permitting electric lines to be run underground, as well as to pipe-in an air supply. This scene was in California Gulch.

RICHARD RONZIO COLLECTION

THE GRANT SMELTER was built at the foot of California Gulch, shown here as it looked in the late 1880's.

RICHARD RONZIO COLLECTION

CURTIS TURBINES were used in the Yak tunnel in California Gulch.

OVERLEAF: This panorama portrays Carbonate Hill as viewed from the downtown district at the southeast edge of Leadville. The grade cutting across this view is the line connecting the Colorado Midland with the Colorado & Southern.

DAVID S. DIGERNESS COLLECTION

DAVID S. DIGERNESS COLLECTION

RICHARD RONZIO COLLECTION

THE RESURRECTION MILL in California Gulch processed ore from the Yak tunnel across the gulch. The Resurrection was served by the Colorado Midland. The CM grade continued on up the gulch to the left, behind the mill.

RICHARD RONZIO COLLECTION

DAVID S. DIGERNESS PHOTO

OPPOSITE: Looking north in California Gulch, the Yak tunnel and mill were in view, circa 1898. D&RG trackage serving the mill can be plainly seen—with boxcars lined up to receive loads of concentrate. In the foreground, a glimpse of the Colorado Midland's line can be observed.

THE YAK MINE AND MILL in California Gulch was one of the big producers of the district. This view of the Yak workings is relatively recent and shows the gulch more-or-less as it appears today. The original boom in California Gulch was based on gold sands and took place in the 1860's; however, this rush of activity was short-lived, and within two years, the gulch was virtually deserted. Little interest was paid to the area until 1875, when the Iron-Silver claim was sold for $40,000 to Levi Leiter, partner of Marshall Field, the well-known Chicago merchant. The gulch immediately took on new life—and Leadville's great silver-lead boom was born.

DAVID S. DIGERNESS COLLECTION

DAVID S. DIGERNESS COLLECTION

A NEW WATER SUPPLY was provided for California Gulch placer claims in 1874, when two South Park miners purchased mining operations in the gulch and constructed a 12-mile ditch to divert water. The source of the water was the Arkansas River.

THE SOUTHWEST SIDE of the Carbonate Camp was industrialized for the most part, as this view of 1910 attests. The long trestle (to the left) belonged to the Colorado Midland and led to the Midland roundhouse (on this page). The Denver & Rio Grande's Moyer Branch curved off to the left at this point—headed toward California Gulch. The viewer is looking across a hazy Arkansas Valley in the distance. In the immediate foreground is narrow-gauge trackage of the Colorado & Southern, which served the Penrose and Bison mines. Just below this line is the dual-gauge trackage of the Colorado Midland and the C&S connection. The CM industrial tracks served the Bohn Shaft.

DAVID S. DIGERNESS PHOTO

THE PUMP SHAFT and tipple of the Minnie mine is shown in this color view. The A. Y. & Minnie mines were developed by Meyer Guggenheim and R. B. Graham, who were making $1,000 a day in clear profit by 1880. By 1888, this mining complex was valued at some $14-million, which permitted Guggenheim and sons to establish the Philadelphia smelter at Pueblo—later sold for $10-million. Many of the famous Leadville fortunes were made possible by the thousands of immigrants who poured into the district from eastern Europe to work in the mines, mills and smelters. A second surge of immigration—largely composed of Mexicans—filled job openings in the smelters. However, even the best of the district's jobs were none too good, considering the over-inflated prices being charged for the necessities of life. The ruthless profiteering by Leadville merchants founded many a fortune. Staple groceries, for example, sold for over four times the going rate in Denver.

OPPOSITE, ABOVE and BELOW: The upper scene on page 317 was in the vicinity of Moyer mine in California Gulch, around 1887. The D&RG spur was still narrow-gauge at the time this view—looking south—was photographed by W. H. Jackson. The Moyer mine (below) was at the end-of-track for the D&RG.

G. B. SANBORN PHOTO
RICHARD RONZIO COLLECTION

W. H. JACKSON PHOTO
RICHARD RONZIO COLLECTION

TROUT FISHING. DR. LAW'S LAKE. EVERGREEN.

DAVID S. DIGERNESS COLLECTION

COLORADO MIDLAND'S end-of-track in the Leadville Mining District looked like this about 1910. The structures in the foreground belonged to the A.Y. & Minnie, while the Moyer mining complex was on Iron Hill (to the left).

DAVID S. DIGERNESS COLLECTION

DAVID S. DIGERNESS COLLECTION

LODE MINING on the slopes of California Gulch proved to be profitable for the Moyer mine (in the foreground), and the A. Y. & Minnie mines (on the hillside to the right). Cribbing was used extensively to keep roads clear of tailings from the mine dumps.

DAVID S. DIGERNESS COLLECTION

MOUNT OF THE HOLY CROSS.

RAILWAY VIEWS IN THE LEADVILLE MINING DISTRICT

G. B. SANBORN PHOTO
STATE HISTORICAL SOCIETY OF COLORADO

DUAL-GAUGE TRACK was common in the Leadville Mining District during the early 1900's. A three-pocket link-and-pin coupler had been mounted on the pilot beam of Colorado & Southern locomotive Number 62, which permitted the engine to switch either standard- or narrow-gauge cars.

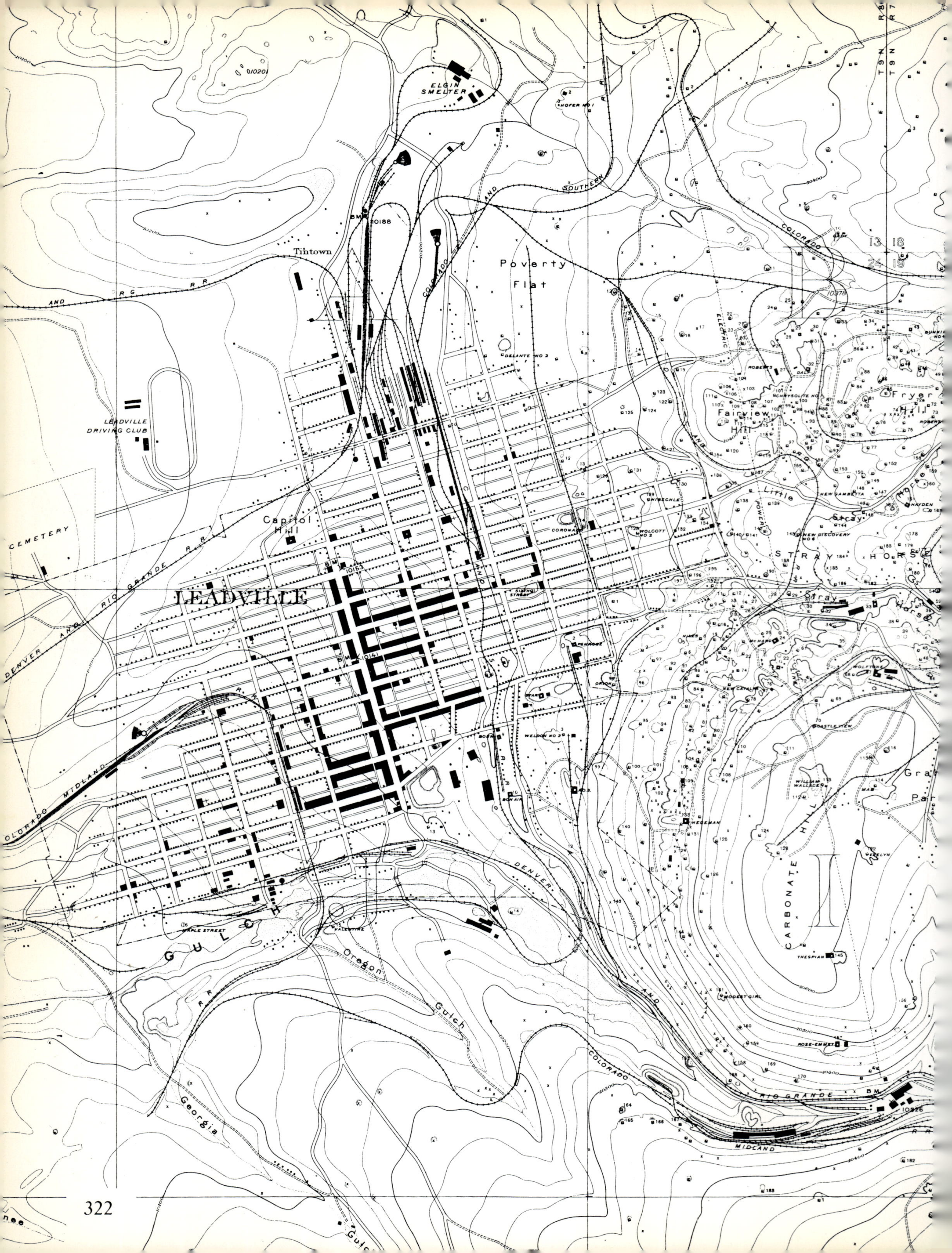
ELGIN SMELTER
Tintown
Poverty Flat
LEADVILLE DRIVING CLUB
CEMETERY
Capitol Hill
LEADVILLE
Fairview Hill
Fryer Hill
Little Stray Horse
STRAY HORSE
CARBONATE HILL
DENVER AND RIO GRANDE R.R.
COLORADO MIDLAND
COLORADO AND SOUTHERN
GULCH
Oregon Gulch
Georgia
RIO GRANDE
MIDLAND

RICHARD RONZIO COLLECTION

THE CORONADO MINE was served by D&RG trackage running down the center of Orange Street. Underground workings were beneath much of the town, as this view taken during the late 1890's indicates.

THE LITTLE JOHNNY MINE was located on Fryer Hill, above Leadville. Metal sheathing had been added to the bucket hoist-tower (to the right).

DAVID S. DIGERNESS COLLECTION

DAVID S. DIGERNESS COLLECTION

THREE MAJOR RAIL LINES served Leadville by 1889: the Colorado Midland, the Colorado & Southern, and the Denver & Rio Grande. This lineup of motive power consisted of older steam engines put to use as switchers in the district. Many of the locomotives—shown here on dual-gauge track—were built in 1880.

THE LEADVILLE ROUNDHOUSE still was narrow-gauge as of June, 1936, when Mac Poor produced this photograph. Colorado & Southern locomotive 69 was resting outside in the morning sunshine beside a pilot plow, wheel sets and an assortment of railroad junk. The photographer was looking toward the north.

M. C. POOR PHOTO
DAVID S. DIGERNESS COLLECTION

DAVID S. DIGERNESS COLLECTION

LEADVILLE'S FIRST RAILWAY DEPOT was built by the Denver & Rio Grande, and it was operated in conjunction with the South Park line under their joint agreement. The depot platform was lined with snow-shovelers waiting for the snow train to take them up the line in order to open up a snow blockade. This photograph was taken during the 1890's, after the Old South Park had moved into their own depot.

A SOUTH PARK TRAIN was photographed during the late 1890's, rattling past the line's depot in Leadville.

CHARLES R. OSGOOD PHOTO
OSGOOD FAMILY COLLECTION

STATE HISTORICAL SOCIETY OF COLORADO

THE UNION PACIFIC built this wooden-frame passenger depot in Leadville for their South Park line. One year after this view was produced—in 1892—the UP's Denver, Leadville & Gunnison moved into their new brick quarters. DL&G Number 162 was assigned switching duty at Leadville during this period. She was originally built as DSP&P Number 35 and was named the "Dillon" in 1882. After C&S took over the line, she became the 22.

THE new brick depot for the South Park line in Leadville was opened in 1893. The South Park shortened their mainline by 20 miles in 1884, with the completion of the "high line" over Boreas and Fremont passes—by way of the Blue and Ten-Mile rivers—between Como, Breckenridge, Dillon and Leadville. The daylight passenger train sported chaircars and was scheduled for 8:45 a.m. departure for Denver.

H. H. BUCKWALTER PHOTO
DAVID S. DIGERNESS COLLECTION

GILBERT AND CASEY PHOTO
STATE HISTORICAL SOCIETY OF COLORADO

THE NORTH END of the South Park's original Leadville depot was photographed in 1892. The depot agent, engineers, firemen, brakemen, friends and the depot hound (on the chair) gathered for this memorable lineup. Number 162 can be seen at the rear of the depot.

THE LAST PASSENGER TRAIN was about to depart from Leadville's C&S depot, April 10, 1937. The train was headed north in order to make its way over Fremont Pass en route to Denver. This depot still stands today, although in rather poor condition.

RICHARD B. JACKSON PHOTO
DAVID S. DIGERNESS COLLECTION

H. H. BUCKWALTER PHOTO
STATE HISTORICAL SOCIETY OF COLORADO

THE COLORADO MIDLAND DEPOT at Leadville was by far the most elaborate in the city. This fancy Victorian structure was located in the western part of the town.

LOADING BOXCARS by sled was not unusual in the Carbonate Camp. Snow was nearly always on the ground during the winter months, and made the movement of wagons most difficult.

DAVID S. DIGERNESS COLLECTION

RICHARD B. JACKSON PHOTO
DAVID S. DIGERNESS COLLECTION

A FAREWELL TO LEADVILLE and the Colorado & Southern narrow-gauge—was the occasion that prompted this photograph to be taken, April 10, 1937. The last passenger train for Denver was nearly ready to depart from the depot, but before leaving town for the last time, the crew took a moment to record this sad event. From left to right: Curley Colligan, Thomas St. John, Roy F. Hight and Pete Marinelli.

LA MAR M. KELLY PHOTO
DAVID S. DIGERNESS COLLECTION

C&S NUMBER 74 was an 1898 product of the Brooks works. Built for the Colorado & North Western (as their Number 30), this handsome 2-8-0 went to the Colorado & Southern in 1921. The locomotive again was sold in 1949 to the Rio Grande Southern (where the C & S number was retained). The 2-8-0 now belongs to the city of Boulder and is on display in a public park.

THE RINGLING BROTHERS circus did not forget the "Cloud City" during the 1890's. One of their private cars was used for their advertising department — and it had arrived in town ahead of the regular circus train carrying the "big top." Located on the outskirts of town, this ornate depot was used jointly by both the South Park line and the Rio Grande. The station tracks were dual-gauge by this time. Notice the omnibus and carriage taxi used for carrying passengers to and from the depot.

CHARLES R. OSGOOD PHOTO
OSGOOD FAMILY COLLECTION

LEADVILLE

RICHARD H. KINDIG PHOTO
DAVID S. DIGERNESS COLLECTION

DURING THE LAST YEARS of the C&S narrow-gauge operations, Leadville was a favorite spot for railfan photographers. The train (above) was recorded by Dick Kindig, while the engine (below) was shot by Bill Gibson.

WILLIAM A. GIBSON PHOTO
DAVID S. DIGERNESS COLLECTION

JOHN W. MAXWELL PHOTO

WHEN THIS PHOTOGRAPH was taken on May 30, 1941, nearly 4 years had passed since narrow-gauge passenger trains had departed from Leadville's Colorado & Southern depot. In this view, Carbonate Hill — where so many of The Mineral Belt's famous silver strikes were made — can be seen behind the depot. C&S rolling stock occupied the trackage alongside the old red-brick building. The handwriting was on the wall for the Old South Park, for in little more than 2 years, the last C&S narrow-gauge train ever would pull into Leadville from Climax, ending slim-gauge operations on this lonely branch forever.

COLORADO & SOUTHERN locomotive Number 9 was found at Leadville one bitter winter day, February 22, 1934, with a passenger train.

OTTO C. PERRY PHOTO
DAVID S. DIGERNESS COLLECTION

DAVID S. DIGERNESS COLLECTION

THE PHOTOGRAPH ABOVE provides one with a view of the eastern part of Leadville, showing the railway terminals in the distance. The C&S depot is just to the left of the D&RG terminal. The photographer was looking down from Carbonate Hill, circa 1910. The scene on the right-hand page was taken from the same spot—looking off to the north—showing Stray Horse Gulch. The D&RG railway shops and the C&S roundhouse are to the far left.

DAVID S. DIGERNESS COLLECTION

DAVID S. DIGERNESS COLLECTION

C&S ENGINE 76 (on the opposite page) and Number 58 (below) were photographed in Leadville during the last days of narrow-gauge operations on the Old South Park.

DAVID S. DIGERNESS COLLECTION

DAVID S. DIGERNESS PHOTO

A D&RG SECTION-HOUSE stood beside U.S. Highway 24 and State Highway 91 in Leadville on the southwest corner of the Rio Grande's rail crossing. This important grade crossing required warning signals. The author took this view while facing toward the southwest, with the Sawatch Range in the background. It was springtime in the Rockies, but snow lingers late at this elevation.

THE LOCOMOTIVE BELOW was resting in front of the Leadville roundhouse when captured on film. Number 69 was one of eight Baldwin 2-8-0's built for the Denver, Leadville & Gunnison in 1890. She went to the White Pass & Yukon Route in 1943 to help move war-time traffic between Skagway and Whitehorse.

DAVID S. DIGERNESS COLLECTION

MORRISON A. SMITH PHOTO
JOHN W. MAXWELL COLLECTION

A DEEP-BLUE SKY — typical of Colorado's high country — greeted the photographer on the morning of July 2, 1941. Colorado & Southern 2-8-0 Number 75 was in the yard near the depot. She was on the dual-gauge line originally placed there to facilitate the movement of standard-gauge Colorado Midland and Rio Grande equipment over Mineral Belt Railway trackage.

NUMBER 6 originally was a passenger engine; however, in later years she was used in combination with other C&S slim-gauge locomotives to haul freight.

DAVID S. DIGERNESS COLLECTION

RICHARD RONZIO COLLECTION

THE COLORADO MIDLAND YARD was being shoveled out after the record snowfall of 1899. Many men with strong backs carried out the task that bulldozers would handle in today's often wasteful technocratic society. Notice the old-time stub switch near the foreground.

DAVID S. DIGERNESS COLLECTION

SAMPLING BULLION.

H. H. BUCKWALTER PHOTO
STATE HISTORICAL SOCIETY OF COLORADO

THE COLORADO MIDLAND had team tracks featuring loading docks during the 1890's at Leadville. Freighters had a going business delivering goods around the city from the railroad.

DON WILLIAMS COLLECTION

THE C&S ENGINE below was switching freight cars in the Leadville district. The crew members obviously were delighted that someone was taking their picture.

JOHN W. MAXWELL PHOTO

THE SNOWPACK was still piled deep on Colorado's highest peaks — Mount Elbert (left) and Mount Massive (right) — when this photograph was recorded on May 30, 1941. In this morning scene, C&S locomotive Number 75 was heading out of Leadville, bound for Climax with a train of narrow-gauge gondolas. Both standard- and narrow-gauge cars are in evidence in the freight yard.

THE LEADVILLE MINING DISTRICT is one of the most heavily mineralized regions in the world. Some of the underground workings were built directly under the town, which was the reason why railway spurs abounded within the town limits. In the scene below, narrow-gauge gondolas were being loaded with ore for shipment to smelters.

DAVID S. DIGERNESS COLLECTION

RICHARD H. KINDIG PHOTO
DAVID S. DIGERNESS COLLECTION

C&S NUMBER 69 (above) served the Old South Park until the end of narrow-gauge operations, whereas Number 10 lasted until only 1934. As stated previously, Number 69 was a Baldwin 2-8-0 built in 1890, while the 10 spot was turned out by Cooke in 1884.

RICHARD B. JACKSON PHOTO
DAVID S. DIGERNESS COLLECTION

OVERLEAF: Leadville was known as the second greatest silver district in Colorado. The mining camp is popularly known as the "Cloud City," as it is two miles high and often cloud-covered. The Colorado & Southern Railway depot is just in front of the tall Catholic Church spire. The Newell tunnel tailings-dump trestle is in the foreground of this early 1900 scene.

L. C. McCLURE PHOTO
DAVID S. DIGERNESS COLLECTION

DAVID S. DIGERNESS COLLECTION

THIS PRE-1910 SCENE shows a typical Leadville mining operation. Included were: shafthouse, surface facilities and a railway-car loading area. The dual-gauge track was known as the Leadville Mineral Belt Railway, which later was purchased by the C&S—in 1900—to be run as a branch-line of their's.

THE TEAM TRACKS usually were a scene of activity as teamsters arrived with wagonloads of ore from the mines above Leadville. The back-breaking job of shoveling was all in a day's work, as the loads were transferred from wagons to narrow-gauge gondolas for shipment to smelters. Notice the link-and-pin coupler on the gondola to the right.

CHARLES R. OSGOOD PHOTO
OSGOOD FAMILY COLLECTION

CHARLES R. OSGOOD PHOTO
OSGOOD FAMILY COLLECTION

TWO YOUNG LADIES got their picture taken while standing on an end-platform of a D&RG four-wheel "bobber." Perhaps the rock balanced on the gondola was a prized chunk of high-grade ore from one of the Leadville district mines.

CHARLES R. OSGOOD PHOTO
OSGOOD FAMILY COLLECTION

A FINE ASSORTMENT of D&RG boxcars and gondolas line the Leadville team tracks in this late 1890 view. To the far left, it looks as though a locomotive was switching

a boxcar out of a repair shed. Several wagonloads of ore from Ibex Hill were about to be loaded into gondolas.

M. C. POOR PHOTO
DAVID S. DIGERNESS COLLECTION

THE COLORADO & SOUTHERN YARD at Leadville still contained dual-gauge track in June of 1937. Mount Elbert is the peak over the car tops.

IT WAS INDEED A SAD DAY in Leadville when the Lake County Courthouse burned to the ground in 1943. Today, a new brick and glass structure — typical of modern architectural styles — occupies the site as the new courthouse. The old hardware building still stands on the next block.

STATE HISTORICAL SOCIETY OF COLORADO

THE CRYSTAL CARNIVAL

Ticket office — entrance

STATE HISTORICAL SOCIETY OF COLORADO

THE CRYSTAL CARNIVAL
During the winter of 1895-'96, Leadville went all out for one of the most extravagant carnivals put on anywhere in Colorado. This and the following two pages illustrate the event. The main feature of the carnival was the ''Ice Palace'' — built entirely of ice to attract attention to Leadville. The Cloud City's winter show was billed as the ''Crystal Carnival'' and gained national fame for the mining town until the castellated structure melted away during July of 1896.

DENVER PUBLIC LIBRARY — WESTERN HISTORY DEPARTMENT

Statue and entrance

STATE HISTORICAL SOCIETY OF COLORADO

Artist's conception

STATE HISTORICAL SOCIETY OF COLORADO

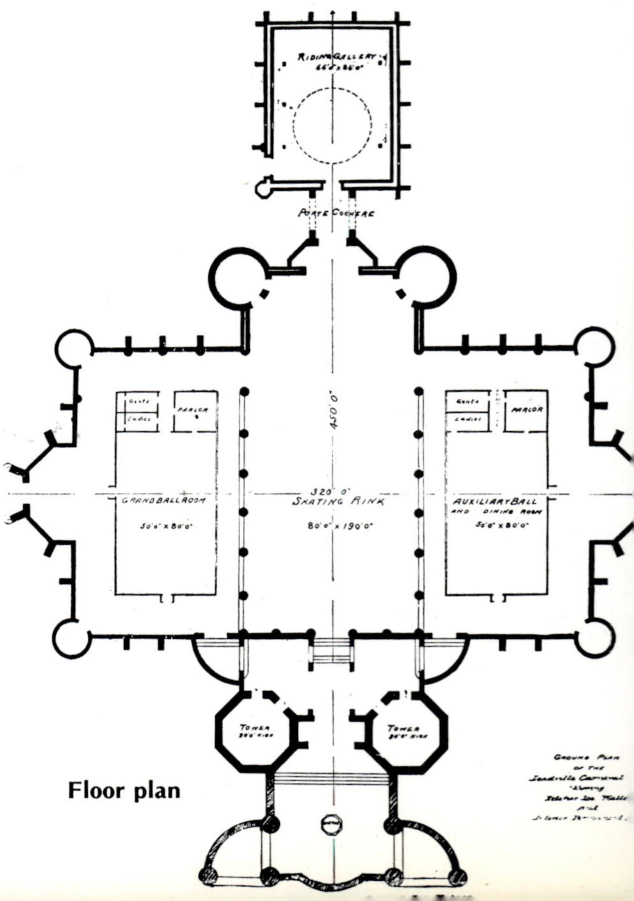

DENVER PUBLIC LIBRARY,
WESTERN HISTORY DEPARTMENT

Ice-skating rink indoors

Floor plan

W. H. Jackson photo

DAVID S. DIGERNESS COLLECTION

DENVER PUBLIC LIBRARY, WESTERN HISTORY DEPARTMENT

Main entranceway

DENVER PUBLIC LIBRARY – WESTERN HISTORY DEPARTMENT

East entrance, featuring Colorado Midland display

STATE HISTORICAL SOCIETY OF COLORADO

THIS VIEW OF LEADVILLE was taken from above Stray Horse Gulch, with the Penn mines in the foreground. This mine was served by trackage of the D&RG off the Chrysolite Extension. The Wolftone and Greenback mines are visible on Carbonate Hill in the background.

PORTFOLIO OF IBEX HILL

DELL A. McCOY PHOTO

PICTURESQUE RUINS of the Big Four mine, including the main shaft building shown here, still stand in the mining region to the west of South Evans Gulch.

DAVID S. DIGERNESS COLLECTION

THE WOLFTONE SHAFTHOUSE and mining properties were on the shoulder of Iron Hill in the Leadville Mining District. Both the D&RG and the C&S narrow-gauge lines served this mine in 1910. The mining cluster on the distant hillside was known as Ibex. Stray Horse Gulch is off to the left.

THE PYRENEES SHAFT on Carbonate Hill was typical of mining structures in the Leadville Mining District. The Wolftone mining complex is virtually hidden by the pithead of the Pyrenees, while the Greenback mine is to the far right.

OPPOSITE: The Graham Park Branch formed part of the maze of trackage in the Leadville district. The D&RG's company photographer, George L. Beam, was looking toward Colorado's highest peaks when he shot this view — Mount Elbert, at 14,433 feet, and Mount Massive, at 14,421 feet. The Wolftone mine can be seen in the right background, and to the left, is the Greenback mine. These mines still were served by the D&RG in 1927. The C&S track cut diagonally under the trestle, ahead of the locomotive.

DAVID S. DIGERNESS COLLECTION

GEORGE L. BEAM PHOTO
JACKSON C. THODE COLLECTION

DAVID S. DIGERNESS PHOTO

THIS COLOR VIEW shows the remains of a snowshed on the Colorado & Southern's spur to the Robert Emmet mine. Even though the precious-mineral lodes of the Leadville Mining District were never exhausted, economic factors caused most of the mines and mills to close—which, in turn, forced the railroads to pull up hundreds of miles of rails in the region.

DAVID S. DIGERNESS COLLECTION

THESE MINERS were out in force at a new mine above Leadville, showing off the latest in raincoats. Underground workings were often very damp. Two wagon-loads of coal had just arrived, along with the superintendent and son, using their own buggy. This was probably near the top of the Ibex Branch before rails reached that point.

DAVID S. DIGERNESS COLLECTION

THE MINERS CO-OPERATIVE UNION, organized by the Knights of Labor, went on strike in Leadville on May 26, 1880. The miners were asking for higher wages and an 8-hour working day. The conflict became so violent that martial law was declared by Governor Pitkin, and the 15-day strike ended on June 10, 1880. Here, a State militiaman stands guard as operations are resumed at the Emmett mine. During the strike, little mining was done and many of the mines filled with water because water pumps were shut down.

DAVID S. DIGERNESS COLLECTION

DAVID S. DIGERNESS COLLECTION

DAVID S. DIGERNESS COLLECTION

ON THIS PAGE are three views of the activity at Camp McIntire. To boost their morale in the frigid cold weather, the militiamen ''took to the air'' or spent the long days just keeping warm inside their tents, cleaning their hand guns and rifles. The only way to keep the men ''on their toes'' was to have daily parade drills.

DAVID S. DIGERNESS COLLECTION

MILITIA REGIMENTAL STAFF and Governor Pitkin posed for the photographer during the miners' strike of 1896. The Governor used his executive power to put down a disturbance between the miners and management in the Leadville district. Below: A cold militiaman stood guard in below-freezing weather at Camp McIntire—just below Leadville, in the Arkansas Valley.

DAVID S. DIGERNESS COLLECTION

DELL A. McCOY PHOTO

DRIFTING CLOUD PATTERNS played among the peaks of the Mosquito Range during a Colorado high-country summer. Traces of D&RG railway grade still exist here—on what what once was the Ibex Branch, high above Leadville. This view shows the point where Alps Junction was located in South Evans Gulch.

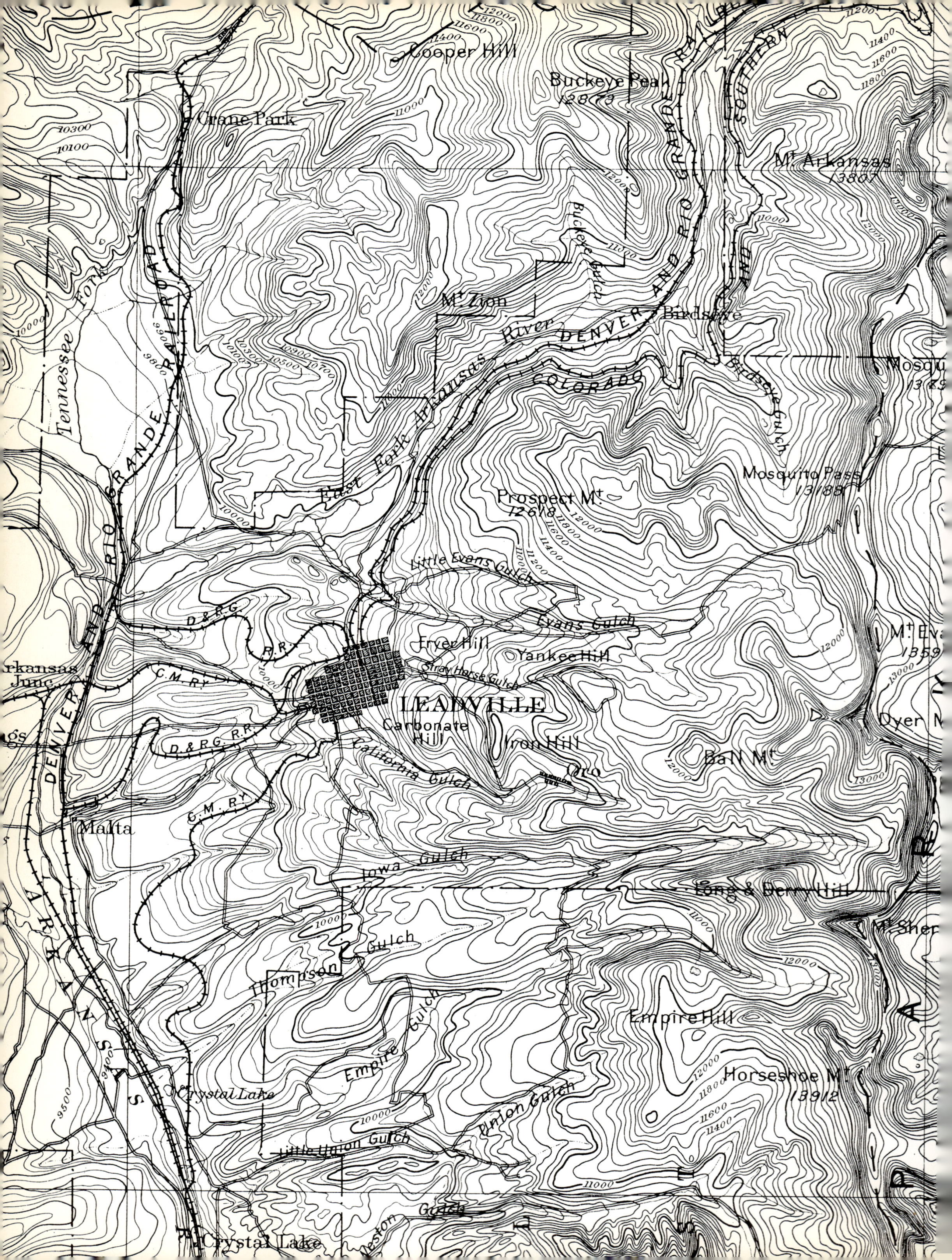

Cooper Hill
Buckeye Peak
12873
Crane Park
Mt Arkansas
13807
Buckeye Gulch
Tennessee Fork
Mt Zion
Birdseye
DENVER AND RIO GRANDE
RAILROAD
SOUTHERN
East Fork Arkansas River
COLORADO
Birdseye Gulch
Mosquito Pass
13188
Prospect Mt
12618
Little Evans Gulch
Evans Gulch
Fryer Hill
Yankee Hill
Stray Horse Gulch
LEADVILLE
Carbonate Hill
Iron Hill
Oro
California Gulch
B&N Mt
D&RG R.R.
C.M. RY
Arkansas Junc.
Malta
Iowa Gulch
Long & Derry Hill
Thompson Gulch
Empire Gulch
Empire Hill
Horseshoe Mt
13912
Crystal Lake
Little Union Gulch
ARKANSAS

RICHARD RONZIO COLLECTION

THIS INTERIOR VIEW shows part of the workings of the Resurrection mill — on the Ibex Branch above Leadville. This mill featured a Wetherell separator and was used for the reduction of gold-and-silver ore from the Ibex Mining District.

MAIL CALL always got a trooper out of his tent, no matter what the weather. These tents were heated with wood or coal, and had the smokestack sticking through the top.

DAVID S. DIGERNESS COLLECTION

DELL A. McCOY PHOTO

ONLY A RELIC of the past, these great old ore bins no longer are served by the D&RG railroad near the top of the world on the Ibex Branch. Located in Lincoln Gulch, the St. Louis Tunnel stands like a sentinel over the rich ore deposits yet to be discovered. This view shows the Mosquito Range in the distance, to the east. Trackage from Penn Junction once climbed past this structure to the right, and it was likely that a spur lead to this tipple.

DAVID S. DIGERNESS COLLECTION

THE IBEX BRANCH — located high above Leadville — was the highest operation of the Denver & Rio Grande. The railway had a continuous battle with winter snowstorms, making a valiant attempt to keep the tracks open each season. This train consisted of a rotary snowplow, two pusher locomotives — numbers 419 and 227 — and one locomotive facing backward to dislodge the train when the front half became stuck in snow — Number 209. Snow-shovelers still were necessary, even with the most modern equipment.

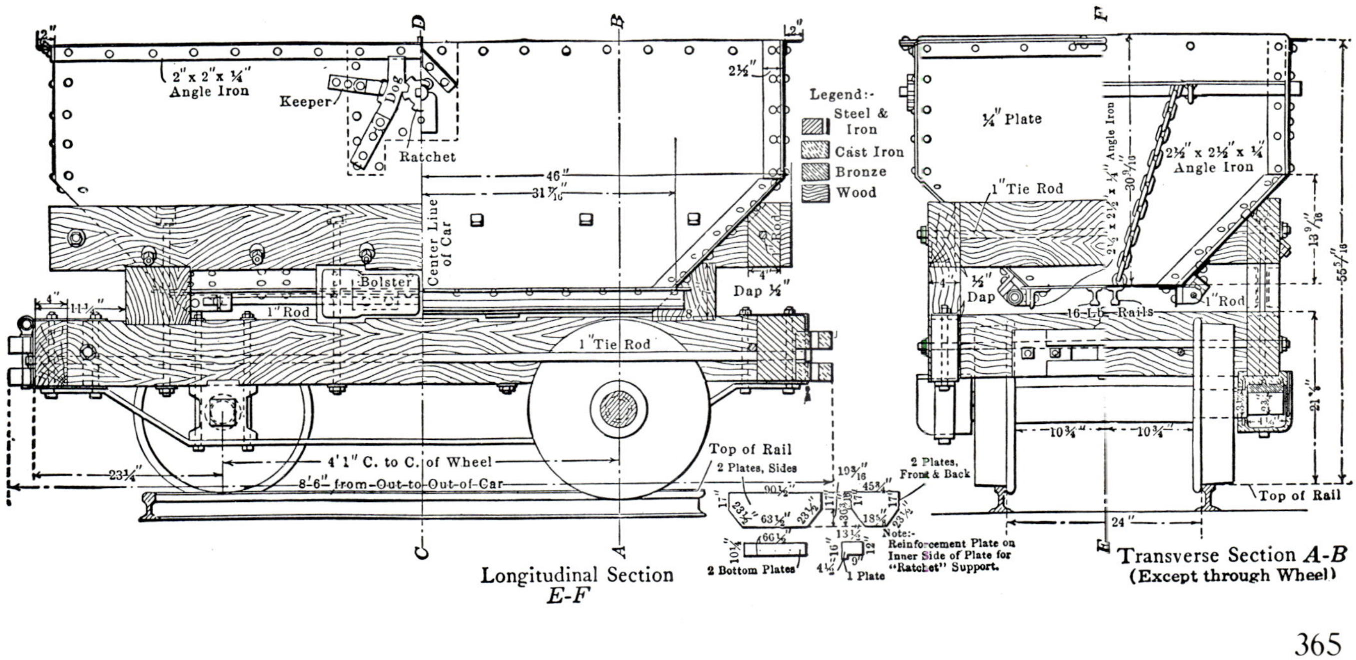

Longitudinal Section *E-F*

Transverse Section *A-B* (Except through Wheel)

RICHARD RONZIO COLLECTION

THE EL PASO MINE was served by the narrow-gauge D&RG Graham Park Branch about the turn of the century. The D&RG's Class 70 2-8-0 locomotives were built in 1881, and originally served as helper engines on mountain passes. Apparently this engine had been demoted to switching service by the time this scene was recorded. In any case, this 2-8-0 had been equipped with

a three-pocket link-and-pin system, which permitted the engine to switch both narrow- and standard-gauge lines in the Leadville district. To the far right, the Wolftone shafthouse and mining complex can be seen on Carbonate Hill.

368 **THE PHOTOGRAPHER** of this view of the Ibex Branch was looking to the northeast from near the D&RG's end-of-track. Resurrection Number 2 was the name of the mine in the distance, where a locomotive was switching the spur. Above this mine is Mosquito Pass, with snow outlining trails leading over it (upper left). The track

RICHARD RONZIO COLLECTION

leading to Resurrection Number 2 is out of sight below the hillside (to the far left). In the right foreground is a mine called the Winnie, with the New Monarch above, served by a spur track. In the upper far right is the Fortune mine, also served by a spur track.

AN ORE WAGON was in the process of being loaded from an ore bunker when this scene was recorded. Wagons of this type were usually hauled by four-horse teams in order to hold back the weight of the ore-load on the descent to Leadville. Ores either were processed in local smelters or shipped out via narrow-gauge gondolas for processing elsewhere. This mine was located on 11,027-foot Iron Hill, high above Leadville. Mount Massive and Turquoise Lake are on the horizon, beyond a smoggy "Cloud City" and above the Arkansas Valley.

CHARLES R. OSGOOD PHOTO
OSGOOD FAMILY COLLECTION

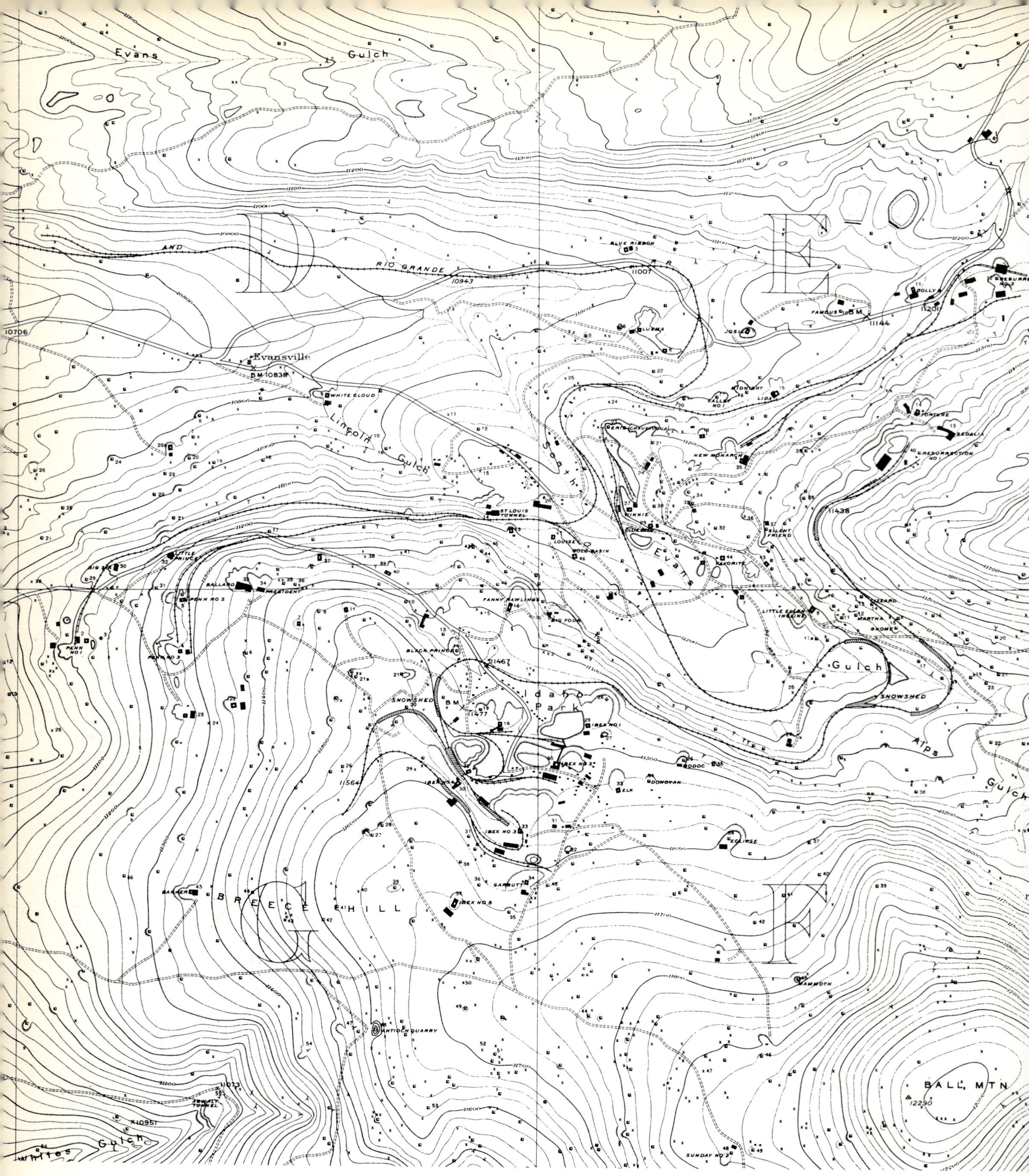

IBEX RAILROAD MAP

DELL A. McCOY PHOTO

SOUTH EVANS GULCH, above Leadville, was the location of many rich mines, including the Winnie, which was served by trackage of the D&RG. The camera was aimed toward the Sawatch Range in the far distance. The old stage road over Mosquito Pass came down through this gulch.

 THIS PANORAMA shows Evans Gulch on the Ibex Branch from the same direction as the previous view, but from a greater distance. The Mosquito Range is in the background. A small settlement of mine workers' homes can be seen on the lefthand page. In the center of the

RICHARD RONZIO COLLECTION

scene, the top mine is the Fortune, and to the right, Resurrection Number 1 was located at the end-of-track, on the same level as the snowshed over the track. Just below the snowshed is the Little Ellen Incline, where there was a reduction works for crushing ore.

DELL A. McCOY PHOTO

THIS UNUSUAL PHOTO shows the railroad turntable pit, which was in use on the Ibex Branch of the D&RG at the elevation of 11,200 feet. Leadville is in the distance, hidden partially by the pines. Turquois Lake is backed up against the Sawatch Range above the Arkansas Valley, in this view looking toward the west.

TRUST CERTIFICATE

Denver, South Park and Pacific

RAILROAD COMPANY.

NUMBER 1710

1000

FIRST MORTGAGE 7 PER CENT BOND

The Farmers Loan and Trust Company of New York hereby certifies that it has received ONE THOUSAND DOLLARS par value of the First Mortgage 7 per cent. Bonds of the Denver, South Park and Pacific Railroad Company, numbered with November 1888 and all subsequent coupons attached. This deposit is made in trust, subject to the terms of an agreement dated the [illegible] March 1889 and subject to the order of Frederick D. Tappen, William [illegible] and [illegible], the Committee named in the said agreement, or a majority of them or their successors, to which said agreement the successive holders hereof assent by receiving this Certificate. The holder hereof is entitled to receive all the securities, benefits and advantages coming to the depositors of said bonds under said agreement.

This Certificate is transferable by delivery, subject to the terms and conditions of the said agreement.

THE FARMERS' LOAN & TRUST COMPANY OF NEW YORK.

New York 1889 by

VICE PRESIDENT.

NEW YORK BANK NOTE COMPANY

ROBERT A. LeMASSENA COLLECTION

DAVID S. DIGERNESS COLLECTION

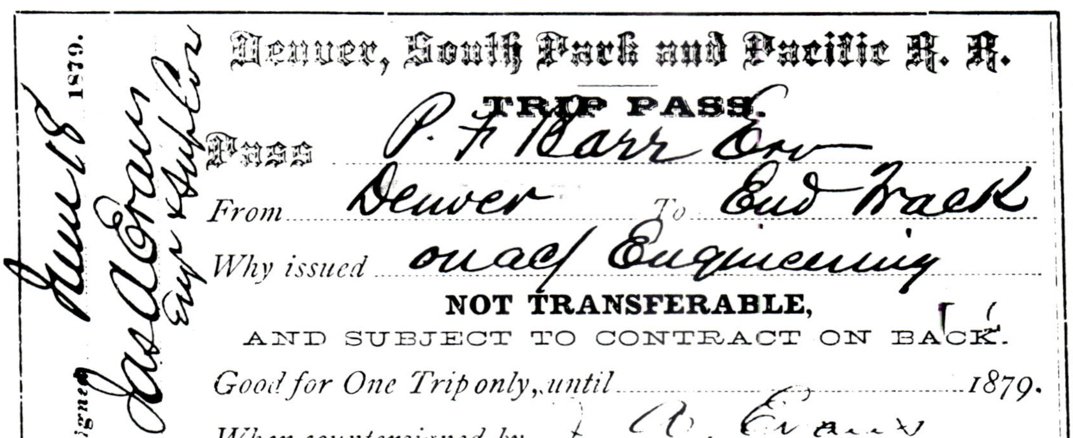

Denver, South Park and Pacific R. R.

TRIP PASS.

Pass P. F. Barr Esq

From Denver To End Track

Why issued on acc Engineering

NOT TRANSFERABLE,

AND SUBJECT TO CONTRACT ON BACK.

Good for One Trip only, until 1879.

When countersigned by J. A. Evans

C. W. Fisher

No. 123

General Superintendent.

June 18 1879.

Countersigned Jas A Evans Eng & Supt Con

THIS WAS IBEX, at the top of the D&RG branch east of Leadville. The depot was just below Ibex Number 4, the mining complex to the left. The trestle behind this complex carried the main branch track, which climbed up from Leadville. Ibex Number 3 appears in the right foreground, and behind it is Ibex Number 2. Ibex Number 1

RICHARD RONZIO COLLECTION

is visible just over the roof of Number 2. The grade with the snowshed, on the far right, led to Resurrection Number 1, which was just out of sight behind the hill. Trails leading over Mosquito Pass are visible in the distance, above two small reservoirs (to the right).

DELL A. McCOY PHOTO

THESE OLD MINE BUILDINGS in Iowa Gulch were still standing about 1970. Numerous similar structures dotted the landscape above Leadville for many years until—one by one—the relentless effects of wind, rain and snow have taken their toll. Today, very little remains as evidence of Leadville's bonanza days.

DELL A. McCOY PHOTO

THE SHAFT HOUSE of the Big Four mining complex is located near South Evans Gulch above Leadville.

 THE HIGH POINT of the Rio Grande's highest rail line was on the Ibex Branch above Leadville — just out of sight, to the far left on Breece Hill — elevation 11,650 feet above sea level. The Ibex depot was listed as 11,520 feet, and is visible on the righthand page, in front of the huge ore-tailings dump. This narrow-gauge seven-mile

RICHARD RONZIO COLLECTION

branch to Ibex was constructed in 1898 and was the highest point ever attained by D&RG rails. Notice the many snowsheds covering the track on both pages. This view looks off to the northwest, across Big Evans Gulch toward the mighty peaks of the Collegiate Range. Lead-

DELL A. McCOY PHOTO

THESE BUILDINGS, part of the Big Four mining complex, were once used as a main shaft house, boarding house, bunk house and blacksmith shop. The Big Four was a precious-metal mine, turning out ore containing gold, silver, zinc and carbonate of lead.

Chapter IX

Condensed History of the

Denver, South Park & Pacific

A LAND OF TOWERING snow-crowned mountains and expansive vistas, Colorado was bypassed by the original transcontinental railroad — Union Pacific. Nevertheless, by the end of the 1890's, a web of rail lines covered the west-central and southwestern regions of the state. The Denver, South Park & Pacific Railroad came into existence in 1873, when the articles of incorporation were submitted to the State of Colorado. Actual construction began in August of the same year, when the Denver Railway Association — a separate construction company — began grading a right-of-way between Denver and Morrison. And by April of 1874, the first DSP&P rails were spiked down on this 16-mile line.

By late June of 1874, the Denver-Morrison line was completed, and service began with two small locomotives: Number 1, a 2-6-0 named the *Fairplay;* and Number 2, a 4-4-0 named the *Platte Canyon.*

After considering the rugged mountain terrain the proposed railway would have to traverse, it was decided to build the Denver, South Park & Pacific as a three-foot narrow-gauge line — even though a track gauge of 4 feet, 8½ inches had been accepted as standard throughout the United States and in most other countries. As a result, the Denver-Morrison line was a "slim-gauge" operation, even though it was entirely east of the Front Range.

The financial depression of 1873 prevented the DSP&P from extending its trackage up the Platte Canyon and into the Colorado Rockies. Economic difficulties caused by this depression brought about a succession of three unsuccessful construction companies. In 1877, however, the Denver & South Park Railroad Construction & Land Company was formed, and building of the railway resumed. As is the case in all depressions, a relatively small number of rich and powerful individuals profited greatly from this economic "crunch," while the overwhelming bulk of the population suffered. As an infant enterprise, the Denver, South Park & Pacific suffered with the rest of the majority and hardly could keep its line to Morrison functioning, much less consider new construction.

As financial conditions improved through 1878, the DSP&P pushed construction from Denver to Bailey, a distance of 56 miles. This line was south of the Morrison route and entered the mountains at Platte Canyon.

The original intention of the DSP&P was to build a line from Denver across South Park to the Arkansas Valley; thence, trackage would go southward into the San Luis Valley to the town of Del Norte. From Del Norte, the DSP&P would strike out "by the most feasible route" to the great San Juan mining districts, and proceed from there to the Pacific Coast. The DSP&P also had projected a route via Summit County and Middle Park to the Pacific. However, neither of these routes was completed.

Before the South Park line reached the Arkansas Valley, a great mining boom took place in the Leadville Mining District — which caused a change in thinking regarding the planned route. During the early part of 1879, the end-of-track was at Webster, at the eastern foot of Kenosha Pass. At this point, a large part of the Leadville freight business was transferred from wagons to DSP&P freight cars, or vice versa. This growing business brought about a management decision to alter the route in order to tap the Leadville riches. As the original goal of reaching the silver-rich San Juans was still viable, the line still was projected to traverse the Arkansas Valley west of Trout Creek — where one line would go north toward Leadville, while the other line would go south toward southwestern Colorado.

As the DSP&P was being extended toward the Arkansas Valley, the Leadville freight business grew to a point that it paid for the construction of new trackage. Meanwhile, during all of this activity, an Eastern capitalist by the name of Jay Gould became interested in the South Park road. This infamous railway manipulator set out to control the DSP&P by late 1879. He already had gained control of the Union Pacific by questionable means and had wormed his way into a half interest in the Denver & Rio Grande.

During February of 1880, the DSP&P reached the banks of the Arkansas River west of Trout Creek Pass. While DSP&P trackage was being pushed west, Jay Gould arranged for a "Joint Operating Agreement" which called for the D&RG to allow the DSP&P to operate trains over D&RG trackage between Buena Vista and Leadville. At the same time, the DSP&P was to permit the D&RG to operate trains over DSP&P trackage between Nathrop and Gunnison via Alpine Pass; however, this part of the agreement was never carried out.

(*Continued on page 396.*)

DAVID S. DIGERNESS COLLECTION

MASON BOGIE Number 55 was built for the DSP&P in 1879 as the 22. She was named the "Crested Butte" and lasted until Denver, Leadville & Gunnison days. Her ultimate disposition is unknown; however, it is known she was in service on the Utah & Northern in 1886.

DAVID S. DIGERNESS COLLECTION

THE COMBINATION baggage, mail and express car (above), and the chaircar (below) are examples of Denver, South Park & Pacific coach stock.

DAVID S. DIGERNESS COLLECTION

DAVID S. DIGERNESS COLLECTION

D&RGW LEASED 2-8-0 Number 343 was in service on the C&S from 1935 until 1937 along with sister engines 345 and 346. These 2-8-0's were fitted with standard C&S cinder-catchers and chutes for their stint on the Old South Park. These engines were needed largely as a result of increased freight shipments to and from the Climax Molybdenum mine atop Fremont Pass.

RED PLUSH UPHOLSTERY, handsome kerosene-lit chandelier lamps and large side windows, which opened to the scent of pine trees and coal smoke; that was Colorado mountain railroading. This coach was typical of C&S narrow-gauge passenger equipment on the Old South Park during the 1930's.

M. C. POOR PHOTO
DAVID S. DIGERNESS COLLECTION

DAVID S. DIGERNESS COLLECTION

ORIGINALLY DSP&P NUMBER 7—the "Gunnison"—this 2-6-6T Mason bogie became the 43 in 1885, the number she retained until the Denver, Leadville & Gunnison scrapped her in 1890.

UNION PACIFIC RAILWAY,

COLORADO DIVISION.

SUPPLEMENT No. 1,

TO EMPLOYES' TIME TABLE NO. 21.

To take Effect Sunday, August 1st, 1889.

At 12.05 A. M. Mountain Time.

DESTROY ALL TIME TABLES OF PREVIOUS DATES.

Subject to the General Rules and Regulations annexed to General Time Table for Colorado Division. Important Changes have been made which must be understood alike by all.

THIS TIME TABLE is for the use and guidance of Employes only, and is not intended for the information of the public, or as an advertisement of time of any Train. The Company reserves the right to vary from it at pleasure.

J. B. GILCHRIST,
Assistant Superintendent,
DENVER, COLO.

J. K. CHOATE,
Superintendent,
DENVER, COLO.

E. DICKINSON,
Assistant Gen'l Manager,
OMAHA, NEB.

NEWS PRINT, DENVER, COLO.

DAVID S. DIGERNESS COLLECTION

THE "DENVER"—locomotive 28—was the last Mason bogie purchased by the DSP&P in 1880. There were only four of this unusual 2-8-6T type of locomotive in use on the DSP&P.

Denver & South Park—CANON DISTRICT—Denver and Como.

WESTWARD

Supplement No. 1 to Time Table No. 21. August 1st, 1889.

Length of Sidings in feet	Location of Scales, Water, Fuel and Turn. Stat'ns.	Third Class 415 Freight Leave Daily Exc. Sunday	413 Way Frt. Leave Daily	Second Class 499 Time Frt. Leave Daily Exc. Sunday	459 Frt. & Pass. Leave Daily	Distances between Stations.	STATIONS	Distances from Denver	First Class 401 Mail & Ex. Leave Daily	403 Passenger Leave Daily Exc. Sunday	451 Mail & Ex. Leave Daily	453 Passenger Leave Daily	455 Passenger Leave Daily	457 Mail & Ex. Leave Daily
					9.10 am		D N DENVER Ud De		8.00 am	4.45 pm	8.20 am	11.25 am	3.15 m	6.15pm
180,989	S.T. F.W	2.00 pm	6.50 an	7.00 pm	9.25	1.1	N West Denver R	1.1	8.10	4.55	8.30	11.35	3.25	6.25
						0.9	A., T. & S. F. Ry. Crossing / D. T. & G. R. R. Crossing	2.0						
2,193		2.10	7.00	7.10	9.35	1.0	Valverde	3.0	8.20	5.05	8.40	11.45	3.35	6.35
						2.2	Overland Park	5.2						
1,946		2.25	7.15	7.20	9.45	1.4	Mooreville	6.6	8.27*	5.12*	8.47	11.52	3.42	6.42
813	W	2.30	7.20	7.24	9.50 am / 10.00	1.0	D Sheridan Junction Ja	7.6	8.30	5.15	8.50 am	11.55 am	3.45 pm / 4.35	6.45pm
						0.6	D. & R. G. R. R. Crossing	8.2						
987		2.45	7.51 / 8.01	7.35		2.9	Littleton	11.1	8.40	5.24				
1,302		3.15	8.30	7.53		5.6	D Wheatland Wd	16.7	8.52*	5.37				
3,062	Y. W	3.35	8.50 / 9.00 / 9.10	8.05		3.7	D Platte Canon Cn	20.4	9.00	5.45				
144						3.1	Gaylors Spur	23.5						
1,059		3.52	9.35	8.23		1.1	Mill Gulch	24.6	9.13*	5.58*				
263						1.3	Stevens Gulch Spur	25.9						
551		4.05	9.50	8.33		0.9	Deansbury	26.8	9.21*	6.06*				
5,111		4.18 / 4.28	10.15	8.45		2.6	D South Platte Sp	29.4	9.30	6.15*				
1,549	W	4.45	10.35	8.55		2.3	Dome Rock	31.7	9.38*	6.23*				
900		5.05	10.55	9.08		2.9	Dawsons	34.6	9.49*	6.34*				
460						0.4	Stone Spur	35.0						
759		5.15	11.05	9.15		1.0	Park Siding	36.0	9.54*	6.39*				
179						1.0	Granite Spur	37.0						
	W					1.7	Buffalo Tank	38.7						
2,761		5.40	11.30	9.31		0.9	D Buffalo Fo	39.6	10.08	6.53*				
3,015	Y.F.	6.00 / 6.30	11.50am / 12.30pm	9.46		2.8	N Pine Grove Ni	42.4	10.20	7.05				
355						2.7	Mayfield Spur	45.1						
1,077						1.0	Thompsons	46.1	10.35*	7.20*				
360						1.3	Mouats Spur	47.4						
687	W	7.16 / 7.26 / 7.36	1.10	10.16		0.3	Crossons	47.7	10.41	7.26				
1,178		8.00	1.40	10.36		3.9	D Estabrook Rk	51.6	10.57	7.42				
728		8.15	2.00	10.50		3.0	Baileys	54.6	11.06	7.51				
755						2.9	Rouarks Spur	57.5						
1,171	W	8.45	2.34 / 2.44	11.08		1.6	Slaghts	59.1	11.18*	8.03*				
896		9.00	3.05	11.22		2.9	Meadows	62.0	11.28*	8.13				
2,518	F.W	9.25	3.35	11.42		4.1	D Grant Us	66.1	11.42	8.27*				
2,239	W	9.45	4.05	11.59 pm		3.3	D Webster Ca	69.4	11.54 am	8.39				
590		10.15	4.35	12.30 am		4.7	Hoosier	74.1	12.14 pm	8.59				
2,094						0.7	Dakes	74.8	12.15	9.00				
3,162	Y.	10.30 / 10.40	4.55	12.45		1.2	N Kenosha Kn	76.0	12.25	9.10				
3,059	Y. W	11.05	5.25	1.10		5.1	D Jefferson Jn	81.1	12.42	9.27				
540		11.20	5.40	1.23		3.0	Michigan Spur	84.1	12.52*	9.37				
460	Y.					3.3	Coal Branch Junction	87.4	1.03	9.48				
11,863	T.F.W	11.40pm	6.05 pm	1.45 am		0.8	N COMO Mo	88.2	1.05 pm	9.50 pm				
		Arrive Daily Exc. Sunday	Arrive Daily	Arrive Daily Exc. Monday	Arrive Daily				Arrive Daily	Arrive Daily Exc. Sunday	Arrive Daily	Arrive Daily	Arrive Daily	Arrive Daily
		(9.40)	(11.15)	(6.45)	(0.40)		(88.2)		(5.05)	(5.05)	(0.30)	(0.30)	(0.30)	(0.30)

West-bound trains will have absolute right to the track over East-bound trains of the same or inferior class. See Rule 62.

When any train becomes 12 hours late it loses all time table rights. See Rule 63.

No Train or Engine will leave Denver or Como without Special Order or Release Ticket.

(16 C)

DAVID S. DIGERNESS COLLECTION

THE "SAN JUAN," DSP&P locomotive 4 — a Mason bogie 2-6-6T — was brand new when photographed at the Mason works in 1878. In fact, the locomotive was so new that the painter had not yet finished work on the cab door. After the engine arrived in Denver, a Congdon stack replaced the "diamond" stack shown in this view; and, of course, a headlight was added. However, like all Mason-bogie engines, Number 4 was not too successful.

Denver & South Park—CANON DISTRICT—Denver and Como.

EASTWARD

First Class							Supplement No. 1 to Time Table No. 21. August 1st, 1889.		Second Class		Third Class			
456 Mail & Exp.	**454** Passenger.	**452** Passenger.	**450** Mail & Exp.	**404** Passenger	**402** Mail & Exp.	Distances between Stations		Distances from Leadville	**458** Frt. & Pass.	**488** Time Frt.	**414** Way Frt.	**416** Freight		Minimum time for third class trains between stations.
Arrive Daily	Arrive Daily	Arrive Daily	Arrive Daily	Arrive Daily Exc. Sunday.	Arrive Daily		**STATIONS**		Arrive Daily	Arrive Daily Exc. Monday	Arrive Daily	Arrive Daily Exc. Monday		
15 / 5.45pm	12.35pm	**11 25** / 10.30 am	**8 20** / **8.00** am	**9 10** / 8.40 am	**6 15** / 6.00 pm		D N DENVER Ud De	151.1	2.10 pm					
5.35	12.25	10.20	7.50	**8 30**	5.51	1.1	N West Denver R	150.0	**2 00**	5.30 am	5.15 pm	2.20am		.10
						0.9	A., T. & S. F. Ry. Crossing / D., T. & G. R. R. Crossing	149.1						.04
5.25	12.15	10.10	7.40	**8 20**	5.42	1.0	Valverde	148.1	1.50	5.20	5.05 / 4.55	2.10		.05
						2.2	Overland Park	145.9						.10
5.18	12.08	10.03	7.33	8.13	5.35	1.4	Mooreville	144.5	1.35	5.05	4.40	1.55		.14
5 15pm	12.05 pm	**10 00** am	7.30 am / **7 20**	8.10	5.32	1.0	D Sheridan Junction Ja	143.5	1.30pm	5.00	**4 35**	1.50		.05
						0.6	D. & R. G. R. R. Crossing	142.9						.04
				8 01	**5 24**	2.9	Littleton	140.0		4.47	4.20	1.37		.15
				7.48	5.09*	5.6	Wheatland	134.4		4.22	3.55	1.12		.25
				7.39	4.59	3.7	D Platte Canon Cn	130.7		4.05	**3 35**	12.55		.15
						3.1	Gaylors Spur	127.6						.16
				7.25*	4 45*	1.1	Mill Gulch	126.5		3.45	3.10	12.35		.11
						1.3	Stevens' Gulch Spur	125.2						
				7 17*	4.37*	0.9	Deansbury	124.3		3.32	2.55	12.20		.08
				7.08	**4 28**	2.6	D South Platte Sp	121.7		3.17	2.40	12.05 am		.11
				7.00*	4.20*	2.3	Dome Rock	119.4		3.03	2.10	11.50 pm		.13
				6.49*	4.09*	2.9	Dawsons	116.5		2.45	1.50	11.35		.18
						0.4	Stone Spur	116.1						.02
				6.44*	4.04*	1.0	Park Siding	115.1		2.35	1.40	11.25		.06
						1.0	Granite Spur	114.1						.05
						1.7	Buffalo Tank	112.4						.08
				6.32	3.52	0.9	D Buffalo Fo	111.5		2.15	1.15	11.05		.02
				6.22	3.42	2.8	N Pine Grove Ni	108.7		2.00	12 50 / **12 30**pm	10.50		.15
						2.7	Mayfield Spur	106.0						.12
				6.09*	3.29*	1.0	Thompsons	105.0						.06
						1.3	Mouats Spur	103.7						.07
				6.04	3.24*	0.3	Crossons	103.4		1 28	11.48 am	**10 16** / 10.06		.01
				5.49	3.09	3.9	D Estabrook Rk	99.5		1.03	11.23	9.35		.20
				5.39	2.59	3.0	Baileys	96.5		12.47	**11 06** / 10.56	9.15		.15
						2.9	Rouarks Spur	93.6						.15
				5.24*	**2 44**	1.6	Slaghts	92.0		12.22	10.20	**8 45**		.10
				5.14*	2.34*	2.9	Meadows	89.1		12.05 am	10.00	**8 13** / 8.03		.17
				5.00	2.20	4.1	N Grant Us	85.0		**11 42** pm	9.30	7.35		.24
				4.49	2.09	3.3	D Webster Ca	81.7		11.25	9.10	7.15		.19
				4.30*	1.50*	4.7	Hoosier	77.0		10.55	8.45	6.40		.26
				4.29	1.49	0.7	Dakes	76.3						.02
				4.23	1.43	1.2	N Kenosha Kn	75.1		**10 40**	8.30	6.25		.12
				4.04	1.24	5.1	D Jefferson Jn	70.0		9.27 / 9.17	7.40	**5 25**		.29
				3.54*	1.14	3.0	Michigan Spur	67.0		9.00	7.25	5.10		.14
				3.43	**1 03**	3.3	Coal Branch Junction	63.7						.15
				3.40 am	1.00pm	0.8	N COMO Mo	62.9		8.40 pm	7.00 am	4.45pm		.05
Leave Daily	Leave Daily	Leave Daily	Leave Daily	Leave Daily Exc. Sunday.	Leave Daily				Leave Daily	Leave Daily Exc. Sunday	Leave Daily	Leave Daily Exc. Sunday		
(0.30)	(0.30)	(0.30)	(0.30)	(5.00)	(5.00)		(88.2)		(0.40)	(8.50)	(10.15)	(9.35)		

West-bound trains will have absolute right to the track over East-bound trains of the same or inferior class. See Rule 62.

When any train becomes 12 hours late it loses all time table rights. See Rule 63.

No Train or Engine will leave Denver or Como without Special Order or Release Ticket.

(17 C)

DAVID S. DIGERNESS COLLECTION

IN AUGUST OF 1883, the Cooke Locomotive Works rolled out Number 63 for her official portrait. The boiler glistened with a burnished Russia-iron jacket and a gun-blue Congdon stack sat atop the smokebox—while gold-leaf lettering and gilt striping set off this fine example of 19th Century craftsmanship. This 2-8-0 30-ton was the mainstay of South Park freight power, capable of hauling over 13,000 pounds.

Denver & South Park—HIGH LINE DISTRICT—Como and Leadville.

Supplement No. 1 to Time Table No. 21. August 1st, 1889.

Length of Sidings in feet, and Location of Scales, Water, Fuel and Turning Stations		WESTWARD 3d Class 417 Freight Leave Daily Exc. Sunday	Second Class 499 Time Frt. Leave Daily Exc. Monday	Second Class 483 Frt. & Pass. Leave Daily Exc. Sunday	First Class 401 Mail & Ex. Leave Daily	Distances from Denver		STATIONS		Distances from Leadville	EASTWARD First Class 402 Mail & Ex. Arrive Daily	Second Class 482 Frt. & Pass. Arrive Daily Exc. Sunday	Second Class 488 Time Frt Arrive Daily Exc. Sunday	3d Class 416 Freight Arrive Daily Exc. Sunday	Minimum time for third class trains between stations
11,863	T.F.W	7.15 am	2.05 am		1.25 pm	88.2	N	COMO	Mo	62.9	12.40 pm		7.40 pm	3.20 pm	
1,734						88.7		0.5 Coal's Spur		62.4					.03
808	W	8.05	2.40*		1.50*	93.8		5.1 Half Way		57.3	12.16*		7.10	2.52	.30
665		8.30	2.57*		2.04*	96.6		2.8 Selkirk Spur		54.5	12.05*pm		6 55	2.32	.17
1,566	T. F.W	8.50	3.10		**2.15**	98.7	D	2.1 Boreas	Bo	52.4	11.55 am		6.40	**2.15** / 2.05	.13
480		8.57	3.16*		2.20*	99.9		1.2 Farnham Spur		51.2	11.50*		6.29	1.50	.07
738		9.05	3.20*		2.23*	100.8		0.9 Dwyer Spur		50.3	11.46*		6.20	1.40	.06
	F.W					102.1		1.3 Bakers Tank		49.0					.08
814		9 25	3.32*		2.35*	103.7		1.6 Argentine		47.4	11.30*		5.55	1.10	.10
322						105.8		2.1 Washington Spur		45.3					.13
565		9.40	3.43*		2.45*	106.4		0.6 Mayo Spur		44.7	11.19*		5.35	12.40	.05
297						108.1		1.7 Suttons Spur		43 0					.08
212						108.5		0.4 Little Mountain Spur		42.6					.03
3 965	Y.F.W	**10.15**	4.00	3.15 pm	3.00	110.0	D	1.5 Breckenridge	Hd	41.1	11.00	**10.15** / 9.40 am	5.05	12.01 pm	.08
175						111.7		1.7 Bartholomews Spur		39.4					.08
1,072		10.37 / **10.47**	4.14*	3.30	3.11*	113.3		1.6 Braddocks		37.8	**10.47***	9.25	4.40	11.30 am	07
4,383	Y.F W	**11.05** / 11.45 am	4.25	3.40 pm / **4.25**	3.22	116.4	D	3.1 Dickey	Jd	34.7	**10.38**	9.15 am	**4 25** / 4.10	11.05 / 10.38 / 10.28	.16
669					3.29	118.3		1.9 Hathaways		32.8	10.31				.11
1,141		12.15 pm	4.42*		3.35*	119.9		1.6 Frisco		31.2	10.25*		3.52	10.05	.09
728		12 30	4.53*		**3 42**	122 0		2.1 Curtin		29.1	10.18*		**3 42** / 3.32	9.50	.13
1,355	W	1.00	5.13*		3.57*	126.1		4.1 Wheeler		25.0	10.05*		3 05	9.30	.20
281						130.5		4.4 Wilders Spur		20.6					.22
	W					132.2		1.7 Kokomo Tank		18.9					.09
1,214		1.55	5.45*		4 20	132.9	D	0.7 Kokomo	Ko	18.2	9.45		2.30	8.55	.05
1,175		2.10 / **2.20**	5.55*		4.27	134.6	D	1.7 Robinson	Rb	16.5	9.38		**2.20**	8.45	.10
1,854	T.F.	2.40	6.10*		4.39	137.4	D	2.8 Climax	Ak	13.7	9.29		2.05	8.30	.15
733		2.48	6.16*		4.44*	138.9		1.5 Alicante		12.2	9.23*		1.55	8.20	.08
297	W					142.3		3.4 French Gulch Spur		8.8					.17
217						143.1		0.8 English Gulch Spur		8.0					.05
532		3.15	6.37*		5.00*	144.8		1.7 Birds Eye Spur		6.3	9.05*		1.20	7.40	.10
	W					147.5		1.7 Three Mile Tank		3.6					.10
						150 4		2.9 D. & R. G. R. R. Crossing		0.7					.15
12 840	S.T. F.W	3.50 pm	**7.00** am		5.20 pm	151.1	D	0.7 LEADVILLE	Vi		8.45 am		12.40 pm	**7.00** am	.04
		Arrive Daily Exc. Sunday	Arrive Daily Exc. Monday	Arrive Daily Exc. Sunday	Arrive Daily						Leave Daily	Leave Daily Exc. Sunday	Leave Daily Exc. Sunday	Leave Daily Exc. Sunday	
		(8.35)	(4.55)	(0.25)	(3.55)			(62 9)			(3.55)	(0.25)	(7.00)	(8.20)	

West-bound trains will have absolute right to the track over East-bound trains of the same or inferior class. See Rule 62.

When any train becomes 12 hours late it loses all Time Table rights. See Rule 63. No Train or Engine will leave Como or Leadville without Special Order or Release Ticket.

(18 C)

DAVID S. DIGERNESS COLLECTION

DSP&P LOCOMOTIVE 35 was fresh out of the Brooks Locomotive Works when this view was produced in 1882. The "Dillon" was later numbered 162 (in 1885) and carried that number until 1899, when she was renumbered 22 by the Colorado & Southern Railway. She served the C&S until 1927, when she was scrapped.

Denver & South Park—ST. ELMO DISTRICT—Como and St. Elmo.

Supplement No. 1 to Time Table No. 21. August 1st, 1889.

Length of Sidings in feet, and	Location of Scales, Water, Fuel and Turning Stations.	WESTWARD 3d Class 441 Freight	2d Class 475 Frt. & Pass.	1st Class 431 Mail & Ex.	Distances from Denver		STATIONS		Distances from Baldwin	EASTWARD 1st Class 432 Mail & Ex.	2d Class 476 Frt. & Pass.	3d Class 442 Freight	Minimum time for third-class trains between stations
		Leave Daily	Leave Daily	Leave Daily						Arrive Daily	Arrive Daily	Arrive Daily	
11,863	S.T.Y.F.W	2.30 am	1.45pm	1.30 pm	88.2	N	COMO	Mo	131.1	12.40pm	12.10pm	5.45 pm	
665		3.00	2.10*	1.42*	93.6		5.4 Red Hill		125.7	12.28*	11.50*am	5 15	.27
920		3.20	2.30*	1.52*	97.8		4.2 Hay Ranch		121.5	12.18*	11.35*	4.55	.20
320		3.45	2.50	2.05*	102.5		4.7 Arthurs		116.8	12.07*	11.15	4 25	.24
3,055	Y.W	3.55	3.00pm	2 10	104.5	D	2.0 Garos	Gn	114.8	12.02pm	11.05 am	4.15	.11
	W				112 8		8.3 Platte River Tank		106.5				.42
3,222		4.40		2.33*	113.6		0.8 Platte River		105.7	11.41*am		3.30	.04
2,228	Y	5.20		**2 50**	119.9	D	6.3 Hill Top	Hq	99.4	11.25*		**2 50** / 2 40	.38
637		5.36		3.01	123.5		3.6 Divide		95.8	11.14		2.15	.13
1,585	W	5.50		3 10*	126.6		3.1 McGees		92 7	11.05*		1.55	.27
1,180		6.15		3.25*	131.5		4.9 Charcoal		87.8	10.49*		1.20	.30
2,634	Y.F.W	6.30		3.30	132.6	D	1.1 Schwanders	Sc	86.7	10.45		1.10	.07
12,646	Y. W	6 45 / 7.05		3.42	136.4	N	3.8 Buena Vista	Bv	90.5	10.32			
							3.8						.20
2,634	Y.F.W	7.30		3.55	132.6	D	Schwanders	Sc	86.7	10 20		12.30	
					135.8		3.2 D. & R. G. R. R. Crossing		83.5				.15
1,019		8.00		4.09*	136.8		1.0 Nathrop		82.5	10.08*		12.05 pm	.05
					141 6		4.8 Haywood Springs		77.7				.30
1,867	Y.F.W	8.35		4.25	142.1		0.5 Hortense		77.2	9.55*		11.35 am	
					144.3		2.2 Raspberry		75.0				13
630		9.05		4.44*	146.9		2.6 Cascade		72.4	9.38*		11.07	.16
838		9.22 / **9 32**		4.52	148.8		1 9 Alpine		70.5	**9 32**		10.55	.12
2,652	T.F.W.	10.00 am / 10.30		5.10 pm	153.3	D	4.5 ST. ELMO	Rs	66.0	9.15 am		**10 30** am	.27
		Arrive Daily	Arrive Daily	Arrive Daily						Leave Daily	Leave Daily	Leave Daily	
		(7.30)	(1.15)	(3.40)			(65.1)			(3.25)	(1.05)	(7.15)	

West-bound trains will have absolute right to the track over East-bound trains of the same or inferior class. See Rule 62.

When any train becomes 12 hours late it loses all Time Table rights. See Rule 63.

No Train or Engine will leave Como or St. Elmo without Special Order or Release Ticket.

Trains moving between Schwanders and Buena Vista will be governed by Buena Vista District card and Rules.

Telegraph signal lamps are not required to be kept burning after regular trains for the day have passed.

(19 C

DAVID S. DIGERNESS COLLECTION

A CLASSIC of DSP&P passenger-train motive power, Cooke-built Mogul Number 71 pulled "varnish" trains for 55 years. Constructed in February of 1884, she later became DL&G Number 113—and, after the C&S took over the road, she was given the number 8. This builder's photograph shows her exquisite original form.

Denver & South Park—GUNNISON DISTRICT—St. Elmo and Baldwin.

Length of Sidings in feet, and Location of Scales, Water, Fuel and Turning Stations.		WESTWARD 3d Class	Second Class		1st Class	Distances from Denver	Supplement No. 1 to Time Table No. 21. August 1st, 1889.	Distances from Baldwin	EASTWARD 1st Class	2d Class	3d Class	Minimum time for third class trains between stations
		441 Freight	437 Frt. & Pass.	435 Frt. & Pass.	431 Mail & Ex.		STATIONS		432 Mail & Ex.	436 Frt. & Pass.	442 Freight	
		Leave Daily	Leave Daily	Leave Daily	Leave Daily				Arrive Daily	Arrive Daily	Arrive Daily	
2,652	T.F.W	**10 20** am			5.10 pm	153.3	D ST. ELMO Rs	66.0	9.15 am		**10 20** am	
							1.7					.10
1,190		10.32			5.18*	155.0	Lady Murphy	64.3	9.09		10.10	
							0.5					.05
1,668		10.40			5.21	155.5	D Romley My	63.8	9.07		10.05	
							2.6					.20
1,852	W	11.05			5.33*	158.1	Hancock	61.2	8.57*		9.40	
							2.9					.25
489		11.25			5.51*	161.0	Atlantic	58.3	8.48*		9.20	
							0.8					.05
660	F. W	11.35			5.54	161.8	D Alpine Tunnel U	57.5	8.45		9.15	
							3.0					.20
500 W ½ M. west		11 50 am			6.09*	164.8	Woodstock	54.5	**8 30**		8.40 / **8.30** / 8.20	
							3.0					.20
276		12.05 pm			6.23*	167.8	Valley Spur	51.5	8.14*		7.55	
							0.9					.05
	W					168.7	Midway Tank	50.6				
							3.2					.15
2,413		12.25			6.40	171.9	Quartz	47.4	7.58		7.15	
							2.8					.15
5,234	Y.F.W	12.40 pm	**3.45** pm		6.52	174.7	D Pitkin P	44.6	7.45	**3.45** / 3.10 pm	6.50 am	
							6.7					.41
1,263	W		4.17		7.15	181.4	Ohio City	37.9	7.25	2.25		
							8.4					.50
1,146	W		5.00		7.40	189.8	Parlins	29.5	7.00	1.30		
							11.4					.57
						201.2	D.& R.G. R.R. Crossing	18.1				
							0.4					.02
						201.6	D. & R. G. R. R. Crossing	17.7				
							0.1					.02
					8.10	201.7	La Veta Hotel	17.6	6 30			
							0.1					
6,996	S.T.Y.F.W		6 00 pm	7.10 am	8.15 pm	201.8	D Gunnison Gu	17.5	6.25 am	12.30 pm / 11 55 am		
							1.9					.10
2,263				7.20		203.7	Gunnison Smelter	15.6		11.45		
							1.3					.07
170				7.30		205.0	Vidals Spur	14.3		11.38		
							4.9					.31
237				8.00		209.9	Teachout	9.4		11.13		
							3.5					.19
307				8.20		213.4	Hinkles	5.9		10.55		
							3.1					.18
3,307	Y.W			8.40		216.5	Castleton	2.8		10 40		
							2.8					.17
2,959	S.F			9.00 am		219.3	BALDWIN			10.20 am		
		Arrive Daily	Arrive Daily	Arrive Daily	Arrive Daily				Leave Daily	Leave Daily	Leave Daily	
		(2.20)	(2.15)	(1.50)	(3.05)		(66.0)		(2.50)	(4.50)	(3.30)	

West-bound trains will have absolute right to the track over East-bound trains of the same or inferior class. See Rule 62.

When any train becomes 12 hours late it loses all Time Table rights. See Rule 63.

No train or engine will leave Gunnison, or St. Elmo without Special Order or Release Ticket.

Telegraph signal lamps are not required to be kept burning after regular trains for the day have passed.

(20 C)

M. C. POOR PHOTO
DAVID S. DIGERNESS COLLECTION

COMBINATION baggage-chaircar 26 was stationed in Leadville during her last years of service—until 1943.

D. & S. P.—MORRISON DISTRICT.

Supplement No. 1 to Time Table No. 21. August 1st, 1889.

Length of Sidings in feet, and Location of Scales, Water, Fuel and Turning Stations.		WESTWARD 2nd Class 459 Frt. & Pass. Leave Daily	First Class 457 Mail & Ex. Leave Daily	455 Passenger Leave Daily	453 Passenger Leave Daily	451 Mail & Ex. Leave Daily	Distances from Sheridan.	STATIONS	Distances from Morrison	EASTWARD First Class 450 Mail & Ex. Arrive Daily	452 Passenger Arrive Daily	454 Passenger Arrive Daily	456 Mail & Ex. Arrive Daily	2nd Class 458 Frt. & Pass. Arrive Daily	Minimum time for third-class trains betw'n stations.
813	W	**10 00** am	6 45 pm	3.45 pm	11.55 am	8.50 am		D SHERIDAN JUNCT Ja	9.5	7.30 am	**10 00** am	12.05 pm	5 15 pm	1.30 pm	
								1.1							
815		10.11	6.49	3.49	**12 01** pm	8.54	1.1	Fort Logan	8.4	7.26	9.56	**12 01** pm	5.11	1.25	.06
								1.9							
205		10.22	6.55	3.55		9.00	3.0	Gillmans	6.5	7.21	9 51		5 06	1.15	.11
								2.1							
250		10.33	7.02	4 02		9.07	5.1	Lee Siding	4.4	7.15	9.45		5.00	1.05	.11
								1.6							
549		10.43	7.07	4.07		9.12	6.7	Mt. Carbon	2.8	7.10	9 40		4.55	12.58	.09
								2.8							
6,694	T. W	11.00 am	7.15 pm	4.15 pm / **4.45**		9 20 am / **9 30**	9.5	D MORRISON Ms		7.00 am	**9 30** am		**4 45** pm	12.45 pm	.15
		Arrive Daily	Arrive Daily	Arrive Daily	Arrive Daily	Arrive Daily				Leave Daily	Leave Daily	Leave Daily	Leave Daily	Leave Daily	
		(1.00)	(0.30)	(0.30)	(0.06)	(0.30)		(9.5)		(0.30)	(0.30)	(0.04)	(0.30)	(0.45)	

Telegraph signal lamps are not required to be kept burning after the regular trains for the day have passed.

D. & S. P.—BUENA VISTA DISTRICT.

Supplement No. 1 to Time Table No. 21. August 1st, 1889.

Length Sidings in ft., and location of Scales, Water, Fuel and Turning Stations		WESTWARD 3d Class 441 Freight Leave Daily	First Class 433 Mail & Ex. Leave Daily	431 Mail & Ex. Leave Daily	Distances from Schwanders	STATIONS	Distances from Buena Vista	EASTWARD First Class 432 Mail & Ex. Arrive Daily	434 Mail & Ex. Arrive Daily	3d Class 444 Freight Arrive Daily	Minimum time for third-class trains betw'n stations.
2,654	F.Y.W	6.30 am	10.20 am	3.30 pm		D SCHWANDERS Sc	3.7	10.45 am	3.55 pm	7.20 am	
						3.4					
					3.4	D. & R. G. R. R. Junct.	0.3				.18
						0.3					
12,646	Y. W	6.45 am / **7 05**	**10 32** am	**3 42** pm	3.7	N BUENA VISTA Bv		**10 32** am	**3 42** pm	**7 05** am	.02
		Arrive Daily	Arrive Daily	Arrive Daily				Leave Daily	Leave Daily	Leave Daily	
		(0.15)	(0.12)	(0.12)		(3.7)		(0.13)	(0.13)	(0.15)	

D. & S. P.—COMO COAL MINE DISTRICT.

Supplement No. 1 to Time Table No. 21. August 1st, 1889.

Length of Sidings in feet, and Location of Scales, Water, Fuel and Turning Stations.		WESTWARD Second Class 463 Frt. & Pass. Leave Daily Exc. Sunday	461 Frt. & Pass. Leave Daily Exc. Sunday	Distances from King	STATIONS	Distances from Coal Branch Junc	EASTWARD Second Class 462 Frt. & Pass. Arrive Daily Exc. Sunday	464 Frt. & Pass. Arrive Daily Exc. Sunday	Minimum time for third-class trains betw'n stations.
6,510	W	3.35 pm	9.35 am		KING	3.3	8.30 am	2.30 pm	
					3.3				
1,705	Y	4.05 pm	10.05 am	3.3	COAL BRANCH JUNC		8.00 am	2.00 pm	.24
		Arrive Daily Exc. Sunday	Arrive Daily Exc. Sunday				Leave Daily Exc. Sunday	Leave Daily Exc. Sunday	
		(0.30)	(0.30)		(3.3)		(0.30)	(0.30)	

West-bound trains will have absolute right to the track over East-bound trains of the same or inferior class. See Rule 62.
When any train becomes 12 hours late it loses all Time Table rights. See Rule 63.

21 C)

WILLIAM A. GIBSON PHOTO
DAVID S. DIGERNESS COLLECTION

C&S NUMBER 30 baggage-chaircar was photographed in Leadville during the last years of narrow-gauge operations.

D. & S. P.—KEYSTONE DISTRICT.

WESTWARD					Supplement No. 1 to Time Table No. 21. August 1st, 1889.	EASTWARD			
Length of Siding in feet, and Location of Scales, Water, Fuel and Turning Stations.		Second Class 483 Frt. & Pass. Leave Daily Exc. Sunday	Second Class 481 Frt. & Pass. Leave Daily Exc. Sunday	Distances from Dickey.	STATIONS	Distances from Keystone.	Second Class 482 Frt. & Pass. Arrive Daily Exc. Sunday	Second Class 484 Frt. & Pass. Arrive Daily Exc. Sunday	Minimum time for third-class trains between stations.
4,383	Y.F.W	3.40 pm			D DICKEY Jd	6.9	9.15 am		
					2.7				.14
2,139		3.55	8.15 am	2.7	D Dillon On	4.2	9.00	4.40 pm	
		4.15 pm }	8.35 am }		4.2				
1,603	Y.	4.20 }	8.40 }	6.9	KEYSTONE		8.40 am	4.20 pm	.22
		Arrive Daily Exc. Sunday	Arrive Daily Exc. Sunday				Leave Daily Exc. Sunday	Leave Daily Exc. Sunday	
		(0.35)	(0.20)		(6.9)		(0.35)	(0.30)	

Telegraph signal lamps are not required to be kept burning after the regular trains for the day have passed.

D. & S. P.—LONDON DISTRICT.

WESTWARD				Supplement No. 1 to Time Table No. 21. August 1st, 1889.	EASTWARD		
Length of Sidings in feet, and location of Scales, Water, Fuel and Turning Stations.		2d Class 475 Frt. & Pass. Leave Daily	Distances from Garos.	STATIONS	Distances from London.	2d Class 476 Frt. & Pass. Arrive Daily	Minimum time for third-class trains between stations.
3,055	Y.F.W	3.00 pm		D GAROS Gn	15.5	11.05 am	
				9.8			.58
1,214			9.8	Lime Rock Track	5.7		
				0.2			.01
2,088	Y.	3.50	10.0	D Fairplay Fp	5.5	10.30	
				5.5			.34
2,745	Y. W	4.25 pm	15.5	D LONDON B		9.55 am	
		Arrive Daily				Leave Daily	
		(1.25)		(15.5)		(1.10)	

Telegraph signal lamps are not required to be kept burning after regular trains for the day have passed.

West-bound trains will have absolute right to the track over East-bound trains of the same or inferior class. See Rule 62.
When any train becomes 12 hours late it loses all Time Table rights. See Rule 63.

(22C)

C. H. KIEFNER PHOTO
DAVID S. DIGERNESS COLLECTION

ORIGINALLY Number 20, this combination baggage-chaircar is now on display in Central City. She was not in regular passenger service when photographed with the number 025 on her sides. This combine was merely 32 feet in length and seated only 14 passengers.

The D&RG completed its line into Leadville during the second half of 1880, and the DSP&P found itself moving a very large volume of freight over this line. In fact, the South Park line had a hard time moving all the freight that came its way!

While the DSP&P was establishing itself as an excellent local narrow-gauge railway, Jay Gould was busy trying every way possible to gain control of the line. Finally, in November of 1880, Gould was able to purchase all the available DSP&P stock, and took over control of both the railway and the construction company. Shortly thereafter, he moved to reduce the South Park road to a mere feeder line of the Union Pacific, and it became the South Park Division of the UP. Thereafter, the interests of the South Park were manipulated by Gould — and *profit* became paramount.

Nevertheless, the South Park was to complete its line into the Gunnison country before construction came to an end. Although the building of the line across the Great Divide by way of Alpine Pass presented the UP with great difficulties,

WILLIAM A. GIBSON PHOTO
DAVID S. DIGERNESS COLLECTION

WILLIAM A. GIBSON PHOTO
DAVID S. DIGERNESS COLLECTION

THE 911 originally was numbered B-2 and was a business car. She was saved by the CB&Q and stored in their Aurora, Illinois, shops.

the South Park was in operation all the way between Denver and Gunnison by late 1882.

Crossing the Continental Divide called for a tunnel some 1,800 feet in length. This bore later became famous as Alpine Tunnel; however, during its construction, it brought the railway nothing but headaches and grief.

After the South Park reached Gunnison, plans were made to extend the line to Utah — and beyond. However, the DSP&P was never to reach beyond the Baldwin coal-mining district northwest of Gunnison, even though some rather expensive grading was done en route to Ohio Pass.

Meanwhile, the South Park was having difficulties with the Rio Grande over the Joint Operating Agreement. This led to the termination of the agreement by the D&RG in February, 1884. The Union Pacific was given prior notice of this termination and proceeded to build a new line from Como to Leadville via Boreas and Fremont passes. At the same time, the UP's South Park line was laying steel from Como toward the Blue River Valley, the Rio Grande was pushing its own rails across Fremont Pass for the valley of the Blue.

Nearly all historians agree that the Union Pacific's control of the DSP&P was very nearly ruinous. And no one can say that the best interests of The Mineral Belt were served properly by the Union Pacific. To make matters worse, a decline in mineral production forced the DSP&P into receivership. In 1889, the road was reorganized as the Denver, Leadville & Gunnison Railway — unfortunately with the Union Pacific still in control of the operation.

In 1890, a consolidation of various Union Pacific lines created a new corporate name for the UP's Colorado and Wyoming operations: The Union Pacific, Denver & Gulf

WILLIAM A. GIBSON PHOTO
DAVID S. DIGERNESS COLLECTION

COACH 75 was built in 1902 by American Car & Foundry. She served the Colorado & Southern until scrapped in 1939. She was 40 feet in length and seated 44.

WILLIAM A. GIBSON PHOTO
DAVID S. DIGERNESS COLLECTION

C&S CAR 089 was rebuilt from a 26-foot freight car of 1872 vintage. Originally she was the B-3, a business car.

Railway. The South Park line was part of this new corporation, but was allowed to keep its own company identity — or what was left of it. The UPD&G lasted until the Silver Panic of 1893 forced the entire Union Pacific into receivership — including the former DSP&P.

During 1898, the Colorado & Southern Railway was formed to operate two Colorado narrow-gauge lines — the South Park and the Colorado Central — as well as over 750 miles of standard-gauge lines in Wyoming, Colorado, New Mexico and Texas. Actual transfer of ownership of the UPD&G lines took place on January 11, 1899. Subsequently, in 1908, the C&S was acquired by the Chicago, Burlington & Quincy Railroad (now the Burlington Northern). And the C&S has operated as a subsidiary of the CB&Q (or BN) since that time.

In 1910, a minor cave-in in Alpine Tunnel gave the new CB&Q-dominated C&S an excuse to abandon this renowned bore under the Great Divide. The poor condition of the tunnel was exaggerated considerably by the railway management in what now is recognized as a ruse to gain quick discontinuation of service over this scenic route.

The line over Trout Creek Pass was abandoned in 1910, also, following wash-outs on this section of track. Previous flood damage to this line always had been repaired; however, the Burlington-controlled C&S obviously was not interested in backing the little South Park line. At the time the CB&Q took over the C&S, the narrow-gauge road had about 412 miles of trackage and generally was considered to be a good local carrier. From that time on, reductions of route mileage became an irreversible company policy until all C&S narrow-gauge lines were either abandoned or converted to standard-gauge.

Without receiving permission to do so, the C&S abruptly discontinued service over the Boreas Pass line in 1910 — between Como and Breckenridge. Legal action on the part of Breckenridge residents and other Coloradans forced the CB&Q subsidiary to reopen the line in 1913.

During 1915, the Colorado & Southern applied for abandonment of the entire South Park line. However, this attempt at abandonment was denied. Not to be thwarted in their negative attempts to abandon the Old South Park, the C&S applied for abandonment several times during the 1920's and 1930's.

Finally, in 1936 — assisted by the effects of the Great Depression — the Interstate Commerce Commission gave the C&S permission to abandon its line between Waterton (near the edge of the Denver metropolitan area) and Climax (14 miles north of Leadville). Abandonment came on April 10, 1937, a truly sad day in the history of Colorado.

One important matter not pointed out to the ICC was the fact that the C&S had so much business during the abandonment deliberations that three Rio Grande engines had to be leased to move all the freight. If this had been brought to the ICC's attention, it is doubtful that the C&S could have gotten rid of its narrow-gauge step-child.

By October of 1938, rails had been removed for the entire distance between Climax and South Platte. Feldspar shipments kept the nine-mile section between South Platte and Waterton in operation until 1941, even though this part of the line technically was abandoned. All of the C&S narrow-gauge trackage in Denver had been standard-gauged by this time, including trackage to Chatfield, which operated on an on-again, off-again basis for several years.

The 14 miles of isolated narrow-gauge road between Climax and Leadville — serving the great Climax molybdenum mine atop Fremont Pass — was converted to standard-gauge on August 25, 1943. This line now is part of the Burlington Northern system, although the corporate name, Colorado & Southern Railway, remains as a relic of the past.

It is indeed too bad that the Burlington's business vision was so myopic, for the Old South Park could have been made into a worthwhile operation, serving a large region in Colorado — if only corporate attitude had been positive instead of so completely negative.

KLAAS GUNNINK COLLECTION

REFRIGERATOR CARS 1118 and 1109 were built by the C&S in 1909. Sister cars 1113, 1116 and 1108 went to the Rio Grande Southern in 1938.

DAVID S. DIGERNESS COLLECTION

DAVID S. DIGERNESS COLLECTION

THE C&S INHERITED its rolling stock in 1899—most of it old and outdated, except for a series of boxcars built by the St. Charles works in 1898. Between 1907 and 1910 the C&S shops duplicated these 30-foot cars, using hardware supplied by American Car & Foundry (successor to St. Charles). By 1909, when the 8226 was built, steel underframes were being included and Bettendorf trucks were standard. The 8226 is shown here as it appeared in 1937, in its last paint scheme.

H. H. BUCKWALTER PHOTO
DAVID S. DIGERNESS COLLECTION

UNION PACIFIC (Denver, Leadville & Gunnison) boxcar 24415 was captured on a glass negative during the 1890's by H. H. Buckwalter. This boxcar is typical of the type of cars the Colorado & Southern obtained when it took over the Old South Park in 1899.

DAVID S. DIGERNESS COLLECTION

BOTH OF THESE GONDOLAS were built in 1902 by the Colorado & Southern. Number 4292 (above) was photographed at Grant in 1937, while the 4308 (below) was at Leadville during the same year. These cars mainly were used to haul coal and molybdenum ore during the last years of the Old South Park.

WILLIAM A. GIBSON PHOTO
DAVID S. DIGERNESS COLLECTION

H. H. BUCKWALTER PHOTO
DAVID S. DIGERNESS COLLECTION

THE DL&G GONDOLA above was classified as a flatcar when photographed by H. H. Buckwalter during the 1890's. Notice the link-and-pin couplers.

DAVID S. DIGERNESS COLLECTION

SITTING IN FRONT of the old Pacific Hotel in Como, Conoco tankcar CONX 35 was part of a Leadville freight train. Oil became an important source of traffic after 1925, when this car was built, and cars such as this one were used on all the inter-connected narrow-gauge lines of Colorado.

WILLIAM A. GIBSON PHOTO
DAVID S. DIGERNESS COLLECTION

PERHAPS THE ODDEST C&S car arrangement was flatcar 1082 with a home-built derrick mounted on it—which was used for bridge and other construction work. A windlass was operated by husky sectionmen in order to power the derrick, a typical example of the way things were done before America became wasteful of energy supplies.

WILLIAM A. GIBSON PHOTO
DAVID S. DIGERNESS COLLECTION

BUILT IN 1890 by the Cooke works as DL&G 011, the 99200 was the first rotary snowplow purchased by the South Park line. Equipped with a new wooden housing and a slope-back tender after being involved in a roundhouse fire in 1935, the plow was later stationed in Leadville.

DAVID S. DIGERNESS COLLECTION

C&S CABOOSE 1000 was a home away from home for the conductor and brakeman assigned to her. Here, at Como—fresh from a run—the crew was airing, or drying, blankets and other items on a makeshift line attached to one of the lamp brackets. Built for the Colorado Central in 1883, Number 1000 was operated by the C&S until the end of the narrow-gauge era.

CONSTRUCTED FOR THE DSP&P in 1883 as waycar 76, this four-wheeled "bobber" was scrapped in 1942. However, through a fortuitous circumstance, the body of the 1009 was saved by the Colorado Railroad Museum and remains as one of only two such cars in existence (the other one—the 1006—being on display at Silver Plume).

DAVID S. DIGERNESS COLLECTION

WILLIAM A. GIBSON PHOTO
DAVID S. DIGERNESS COLLECTION

FOUR-WHEELED "CRUMMY" 1008 was built for the South Park by the Union Pacific in 1884. Unlike similar D&RG four-wheeled bobbers, the C&S never converted its short waycars to double-truck models. Originally built without cupolas, the C&S did add this and other improvements; however, the railway did only what was absolutely necessary. The 1008 was scrapped at Leadville in 1942 after the line to Climax was converted to standard-gauge.

DENVER, SOUTH PARK & PACIFIC

Drawn by DAN ABBOTT — Scale: 3/16″=1′-0″

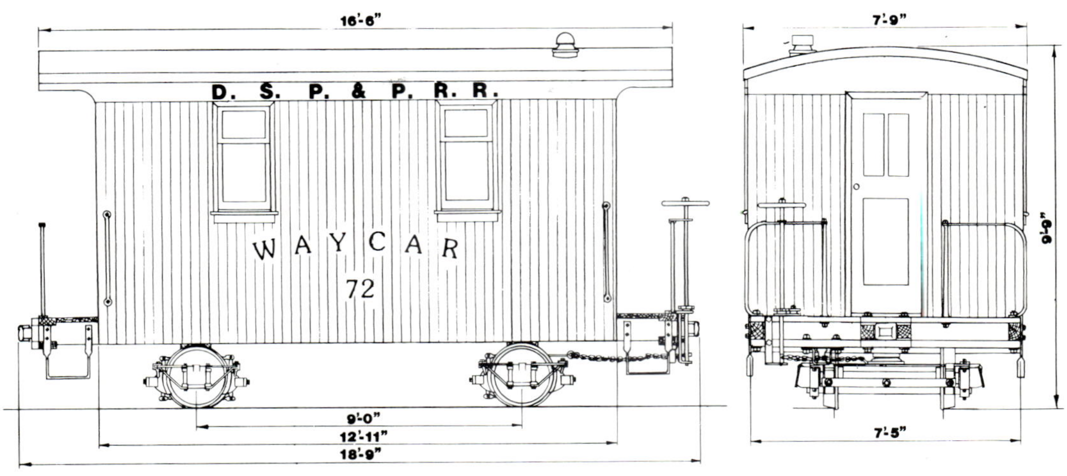

DENVER, SOUTH PARK & PACIFIC

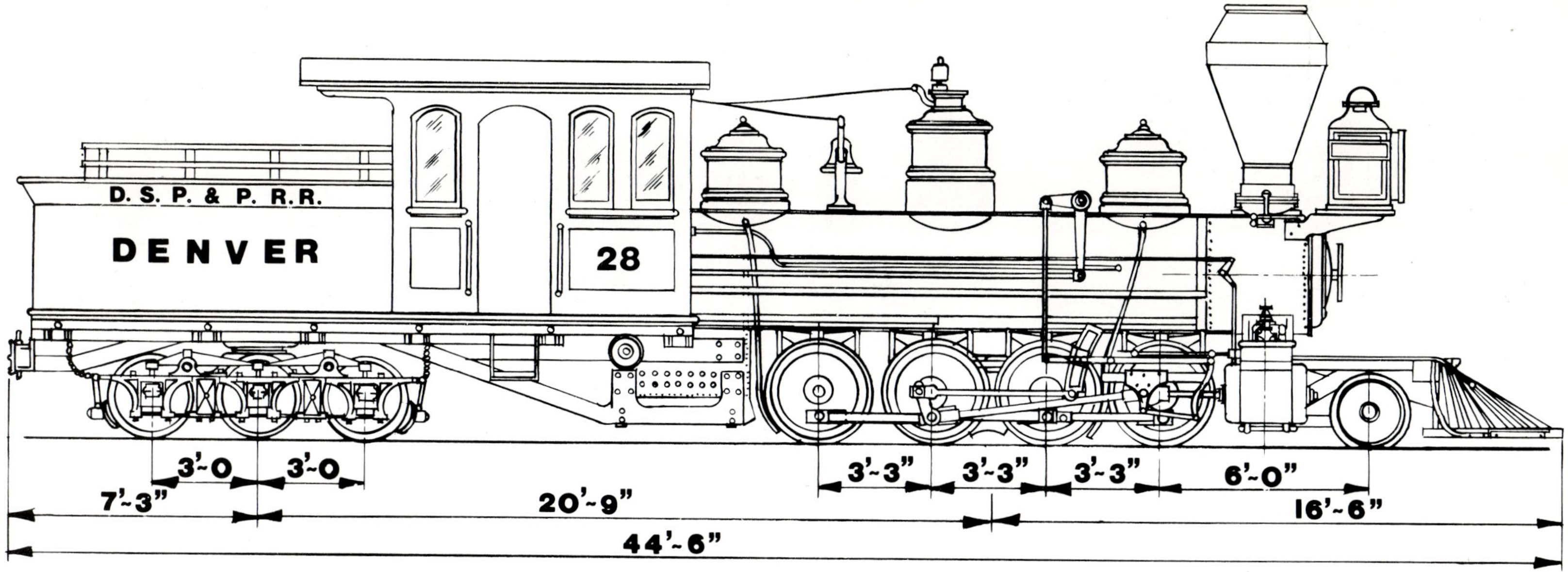

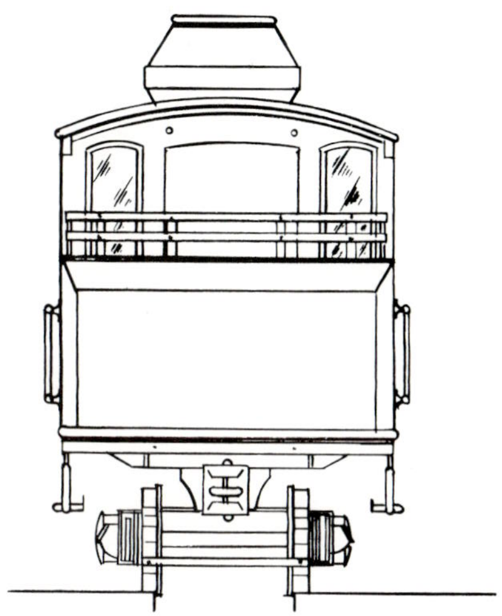

MASON-BOGIE

2-8-6T Number 28

"The Denver"

Drawn by Dan Abbott — Scale: 3/16" = 1'-0"

DENVER, SOUTH PARK & PACIFIC

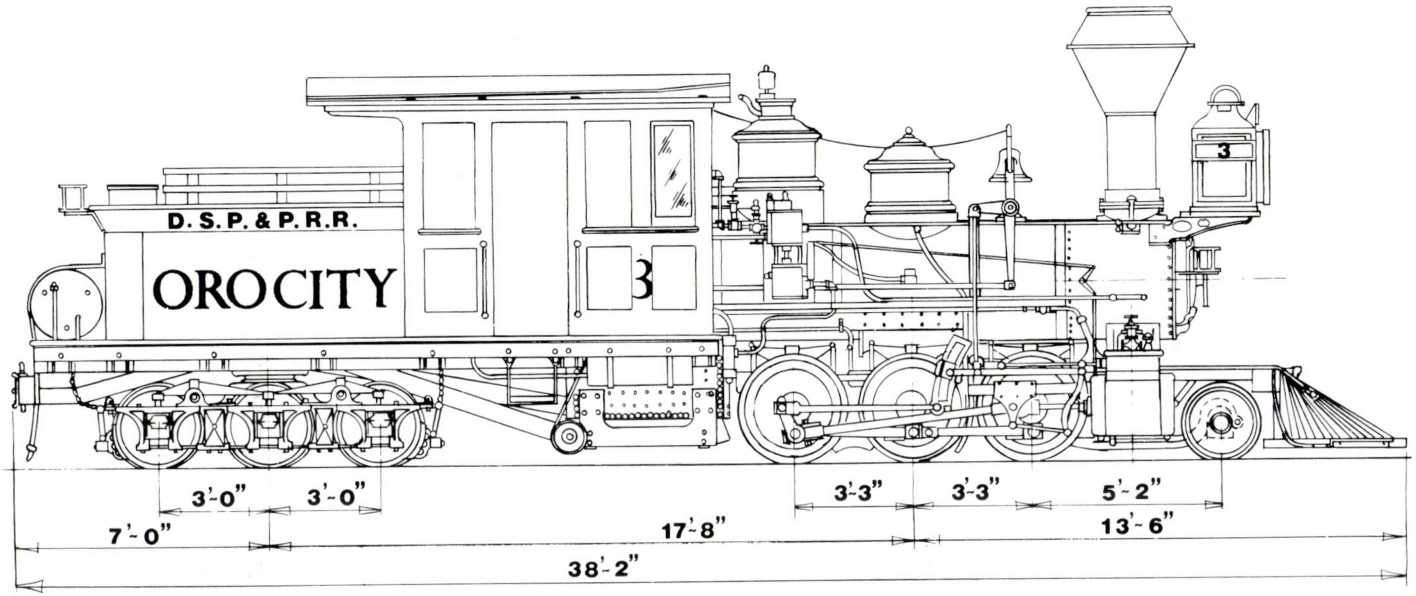

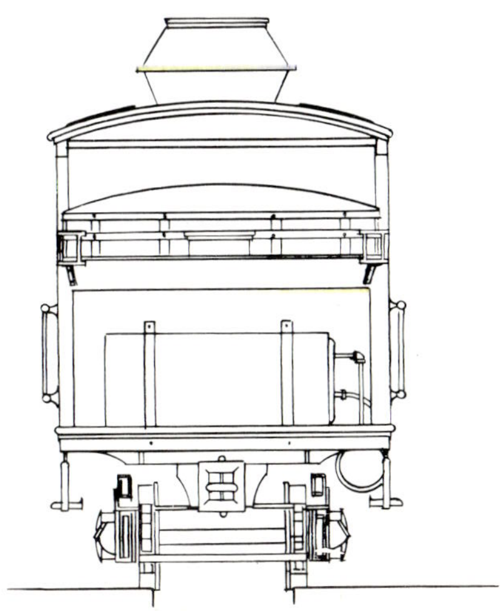

MASON-BOGIE

2-6-6T Number 3

The "Oro City"

Drawn by DAN ABBOTT

Drawn by Dan Abbott — Scale: 3/16" = 1'-0"

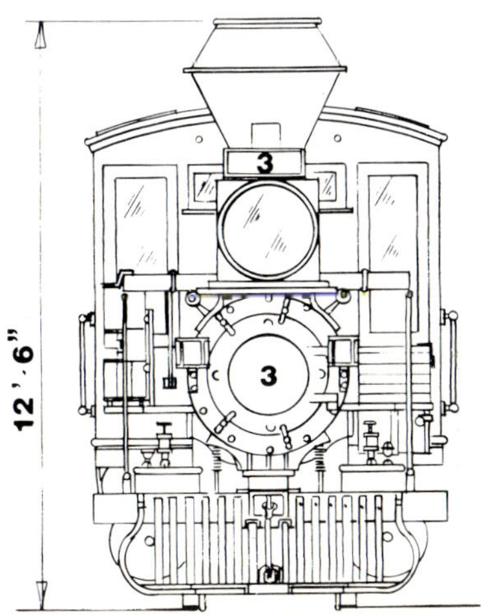

DENVER, SOUTH PARK & PACIFIC

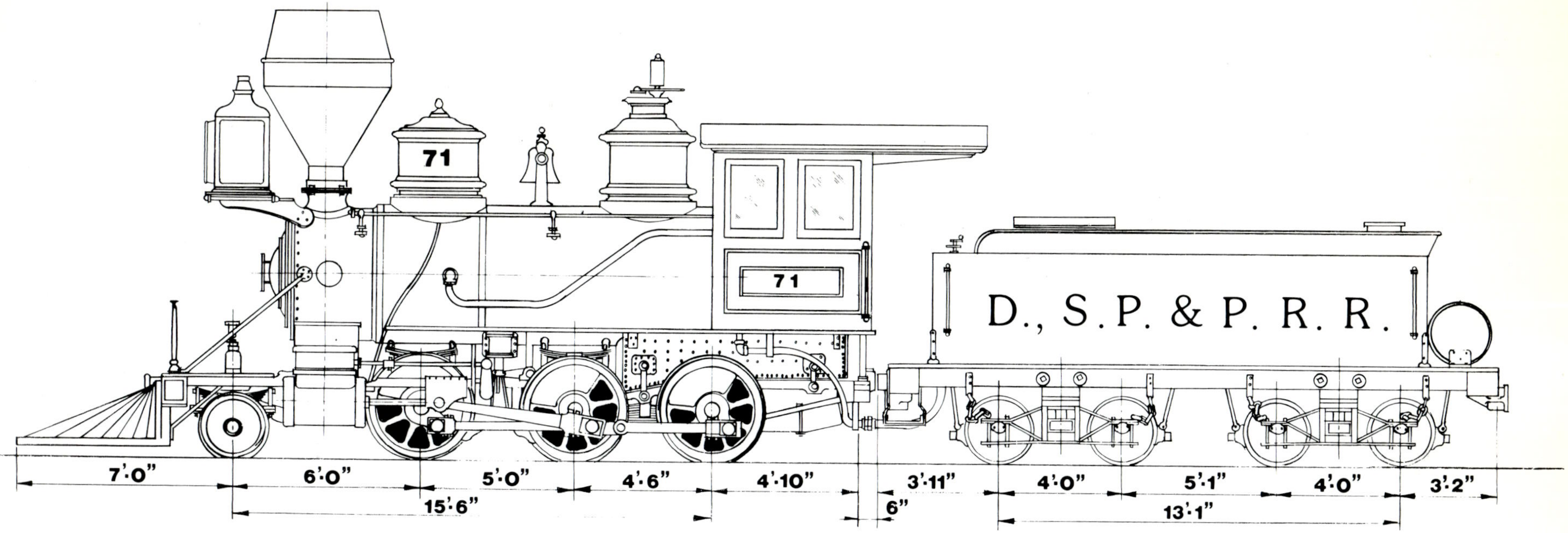

COOKE 2-6-0

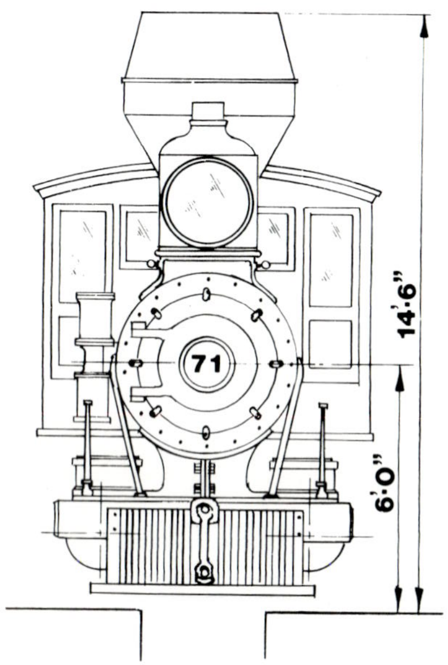

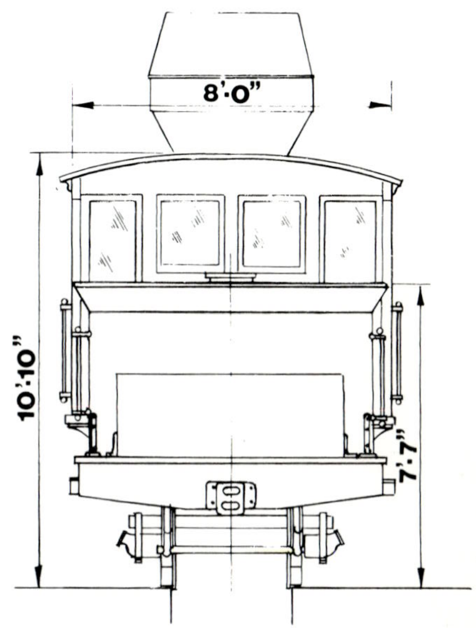

Drawn by Dan Abbott — Scale: 3/16″=1′-0″

DENVER, SOUTH PARK & PACIFIC

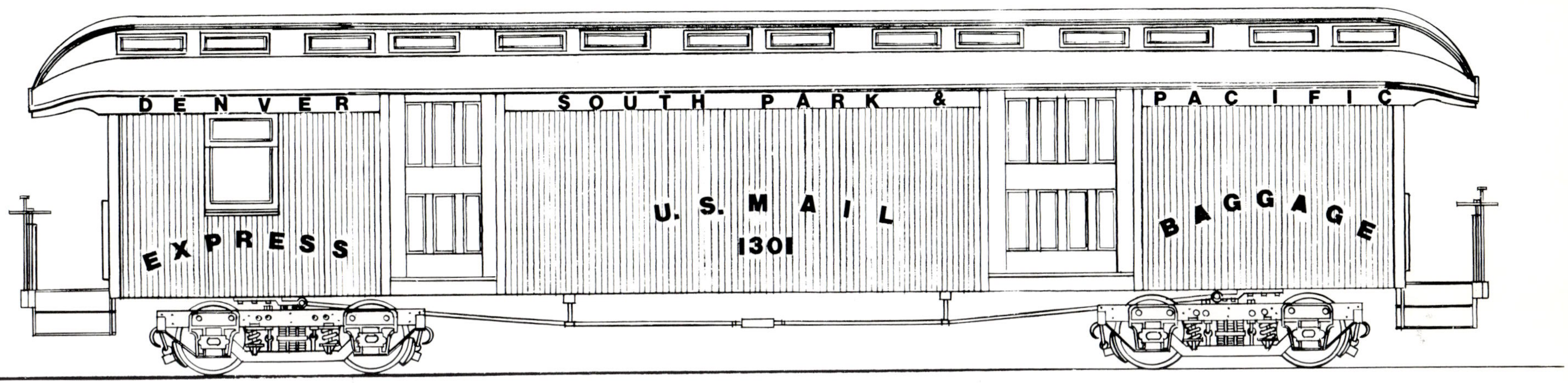

BAGGAGE-EXPRESS-MAIL CAR NUMBER 1301

Drawn by Dan Abbott — Scale: 3/16″=1′-0″

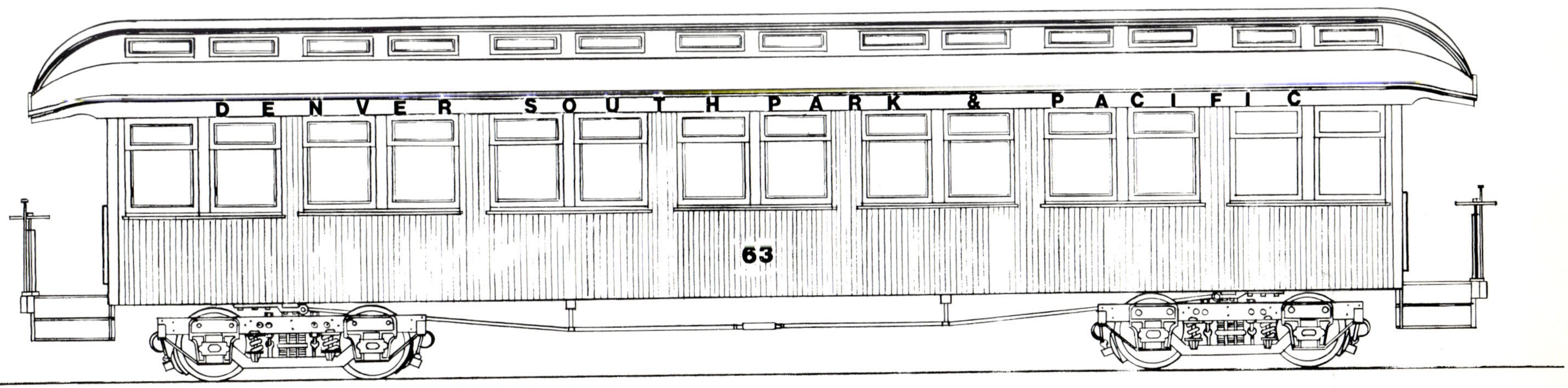

DSP&P CHAIRCAR NUMBER 63

Drawn by Dan Abbott — Scale: 3/16″=1′-0″

DENVER, SOUTH PARK & PACIFIC

TIFFANY SUMMER & WINTER REFRIGERATOR CAR

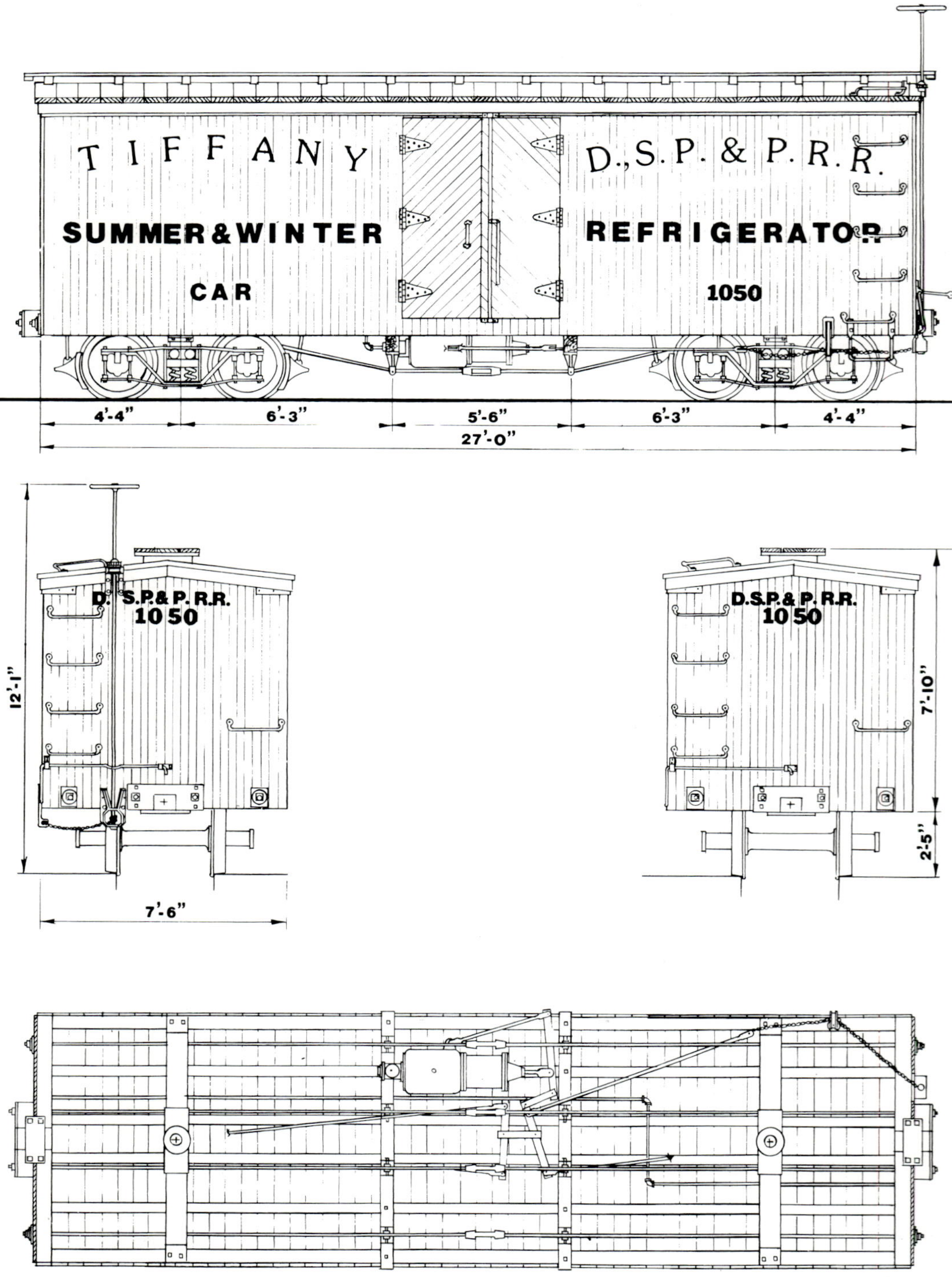

Drawn by Dan Abbott — Scale: 3/16″=1′-0″

INDEX

Asterisk (*) After Page Number Indicates Illustration.

INDEX CONTINUED

INDEX CONTINUED

INDEX CONTINUED

INDEX CONTINUED

INDEX CONTINUED

THE END

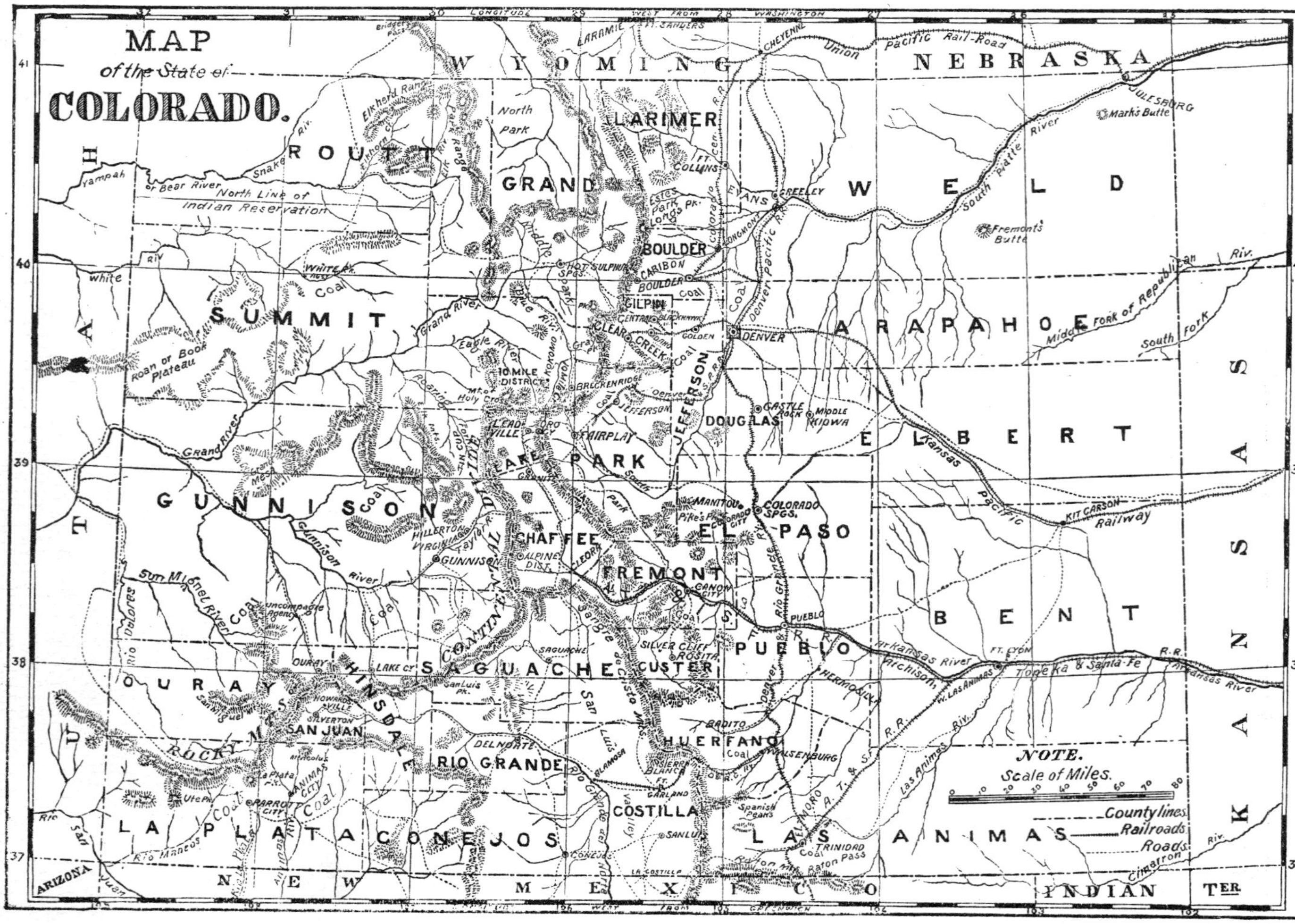
MAP
of the State of
COLORADO.
WYOMING
NEBRASKA
ROUTT
GRAND
LARIMER
WELD
BOULDER
GILPIN
SUMMIT
ARAPAHOE
CLEAR CREEK
JEFFERSON
DOUGLAS
ELBERT
PARK
GUNNISON
CHAFFEE
EL PASO
FREMONT
BENT
PUEBLO
SAGUACHE
CUSTER
OURAY
HINSDALE
SAN JUAN
RIO GRANDE
HUERFANO
COSTILLA
LA PLATA
CONEJOS
LAS ANIMAS
KANSAS
UTAH
NEW MEXICO
INDIAN TER.
ARIZONA
DENVER
CHEYENNE
GREELEY
JULESBURG
KIT CARSON
COLORADO SPGS.
Union Pacific Rail-Road
Kansas Pacific Railway
North Line of Indian Reservation
NOTE.
Scale of Miles.
County lines
Railroads
Roads